T0136862

Mark Dion
Travels of
William Bartram
Reconsidered

HYDRANC A UERCIFOLIA .

Mark Dion

Travels of William Bartram Reconsidered

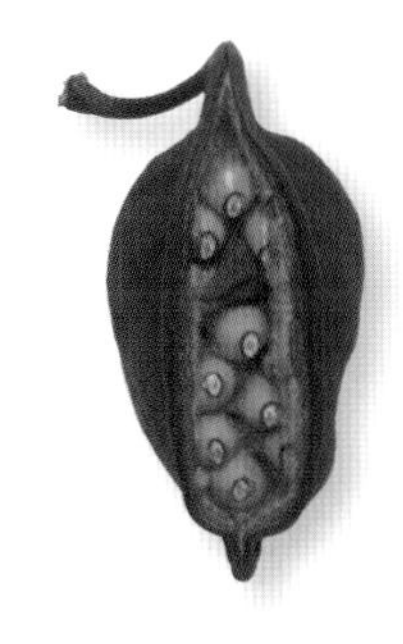

Bartram's Garden
Philadelphia, Pennsylvania

Mark Dion
Travels of William Bartram: Reconsidered
www.markdionsbartramstravels.com

An installation curated by Julie Courtney at
Bartram's Garden
54th and Lindbergh Boulevard
Philadelphia, Pennsylvania 19143
www.bartramsgarden.org

This project has been supported by a grant from the Philadelphia Exhibitions Initiative, and the Marketing Innovation Program, both programs of the Pew Center for Arts and Heritage, funded by The Pew Charitable Trusts, and administered by The University of the Arts, Philadelphia. Additional support has been received from Carole Haas Gravagno and Emilio Gravagno.

Library of Congress Control Number: 2008938331

ISBN: 978-0-615-25748-8

Design: Jeffrey Jenkins

Editing: Joseph N. Newland, Q.E.D.

Table of Contents

Director's Foreword 6

Letters of Introduction 7
Julie Courtney, *Project Curator*
James B. Straw, *President, John Bartram Association*
William M. Lefevre, *Former Executive Director, John Bartram Association*
Joseph P. Riley, Jr., *Mayor of Charleston, SC*
Gregory Volk, *Art Critic and Curator*
Joel T. Fry, *Curator, Bartram's Garden*
Lord Breaulove Swells Whimsy, *Enthusiast and Gentleman Amateur*

Mark Dion's Departure Address 10

Curious Cabinets 12
Julie Courtney

A Short History of Bartram's Garden 18
Joel T. Fry

William Bartram's Travels, 1773–1777 34

William Bartram — A Portfolio 37

Mark Dion Journal Excerpts 48

Mark Dion's *Travels of William Bartram Reconsidered* 66
Gregory Volk
1. **A Cabinet of Alligators** 67
2. **A Cabinet of Plants, a Cabinet of Water** 72
3. **A Cabinet of Hate** 80
4. **Cabinets of Curiosities** 84
5. **A Cabinet of Postcards** 100

About Linnaeus 76

Selected Bibliography 108

Credits 110

Director's Foreword

Projects like Mark Dion's *Travels of William Bartram—Reconsidered* and other collaborative programs at Bartram's Garden—such as concerts with the Ars Nova Workshop and visual artist's installations with the Institute of Contemporary Art—help visitors to better understand the linkages between history and horticulture. While this historic garden is a landscape that in many ways reflects the past, present, and future, it can also be a canvas that allows artists of all disciplines to create their own linkages to nature and help us see the world through new and exciting windows. This enriches the experience at Bartram's Garden and ultimately, enriches and changes lives.

On behalf of the Board of Directors and the staff, we gratefully acknowledge the following individuals and institutions for their participation and support of *Travels of William Bartram—Reconsidered*. To Julie Courtney, guest curator, for her vision of this project; to Dana Sherwood for her creative partnership with the artist; to Bill LeFevre, former Executive Director of Bartram's Garden, for helping to get the project off the ground; to Laris Kreslins and Kendra Gaeta, Lime Projects for their cutting-edge technological support of the exhibition-companion website; and to Lord Whimsy, for giving Mark a wonderful send off. To our lead sponsors, the Philadelphia Exhibitions Initiative with additional support from the Marketing Innovation Program, both programs of the Pew Center for Arts and Heritage, funded by The Pew Charitable Trusts, and administered by The University of the Arts, Philadelphia. To our generous donors: Carole and Emilio Gravagno, SCA Americas, and Citi Smith Barney. To our partners: the American Philosophical Society, the Wagner Free Institute of Science, the Historical Society of Pennsylvania, and the students of Richard Allen Preparatory Charter School. A special thank you to Alina Josan, who became part of the Bartram family and expertly engaged our many exhibition visitors. And finally, our deepest gratitude to Mark Dion for a brilliant exhibition and for setting the stage for innovative programs that open new doors to Bartram's Garden deeper meanings and purpose.

Louise Turan
Executive Director
Bartram's Garden

Julie Courtney
Independent Curator
2227 Bainbridge Street
Philadelphia, PA 19146
215-732-9641
juliescourtney@gmail.com

October 28, 2007

TO WHOM IT MAY CONCERN:

This letter serves as an introduction to the artist, naturalist, archeologist, explorer and all around nice guy who is traveling in your region to retrace the travels of the naturalist and artist, William Bartram (1739-1823) to northern Florida for an art project that will be on view at Bartram's Garden in Philadelphia in June, 2008.

Much in the same way the Bartrams traveled, Mark Dion will meet like-minded people, share ideas, take samples of seeds, earth, water and send the specimens back for display at John Bartram's home in Philadelphia. Like William Bartram, Mark Dion will keep a journal and make paintings and drawings of his discoveries. Unlike William Bartram, Dion's travels will be documented on-line at:

MARKDIONSBARTRAMSTRAVELS.COM

His travels will be mapped with a GPS device and his journals, drawings and records of his encounters will be recorded. This will be a unique opportunity to follow an artist and observe his methods and art making unfold.

I can assure you Mark Dion and his traveling companion, Ms. Dana Sherwood, will enrich your life if you are so kind as to welcome him into your home or institution. They are honest, kind, intelligent and very funny. I know you will enjoy their company.

Most sincerely,

Julie Courtney

Julie Courtney
Curator

November 9, 2007

Gentleperson:

As President of the John Bartram Association Board, it gives me great pleasure to introduce to you renowned artist Mark Dion, who, for reasons known only to him, has decided to retrace the steps of the Bartrams in the southeast United States.

Bartram's Garden and its historic house and outbuildings have recently become host to some of Philadelphia's most notorious artists, and we have learned, therefore, how to deal with people who are decidedly left-brained. Mr. Dion is as left-brained as they come, so I urge you to treat him with the utmost of generosity and loving care.

Word has it that Mr. Dion plans to retrace the exploratory journeys of the Bartrams, in particular, William's expedition to northern Florida. We, at The Garden, are most curious about this focus, since most of the area is now suburban subdivisions and trailer parks. Oh, well.

I'm sure that you will find Mr. Dion pleasant and engaging, and most enjoyable to be with (unlike most artists). His life's work is quite extraordinary and I urge you to surf his new, innovative (that's what good artists do) website at www.MarkDionsBartramsTravels.com.

Thank you, in advance, for your kindness and hospitality toward Mark Dion, and if he gets lost, just send him back home.

With sincerity,

James B Straw

James B. Straw
President

JBS/lb

JOHN BARTRAM ASSOCIATION
Bartram's Garden, 54th St. & Lindbergh Blvd., Philadelphia, Pennsylvania 19143
Phone: (215) 729-5281 · Fax: (215) 729-1047 · www.bartramsgarden.org

November 2007

Dear Sir or Madam,

It is my pleasure to introduce you to Mark Dion and his project, *The Travels of William Bartram – Reconsidered.* In my former capacity as executive director of the John Bartram Association, I was pleased to participate in the planning of this artistic collaboration.

Mr. Dion has recently been artist-in-residence at the Natural History Museum in London, which houses a vast collection of Bartram holdings. I had the privilege of viewing his most recent exhibit there *Systema metropolis* which opened last June to rave reviews. For this project, he will examine the history and culture of 18th century American naturalists, John Bartram (1699-1777) and his son William (1739-1823). Using their journals, maps and drawings, Mr. Dion plans to retrace the exploratory journeys of the Bartrams, in particular, William's expedition to northern Florida. The artist will travel in much the same way as the Bartrams: by horseback, boat, and on foot. He will travel with other "explorers" to collect specimens and man-made artifacts found in this radically altered landscape and, in keeping with his longstanding interest in mail art, send them back to Bartram's Garden.

The artifacts and specimens will be installed in cabinets the artist will design and build for the exhibit, which runs from June to December 2008 in the John Bartram House at Bartram's Garden in Philadelphia. Mark Dion's progress will also be documented on video and uploaded on the project's website: www.markdionsbartramstravels.com.

Your interactions and conversations with Mark will surely influence his art and increase the enjoyment of all who see his work. Our partners include the American Philosophical Society, The Wagner Free Institute of Science, The Academy of Natural Sciences, and the Tyler School of Art. You will find Mark to be highly observant and deeply respectful of the history, culture, and environment he studies for his art. It is my hope that you will find working with him as pleasurable as did I.

Sincerely,

William M. LeFevre
Former Executive Director

John Bartram Association
Bartram's Garden, 54th St. & Lindbergh Blvd., Philadelphia, Pennsylvania 19143
Phone: (215) 729-5281 · Fax: (215) 729-1047 · www.bartramsgarden.org

City of Charleston
Joseph P. Riley, Jr.
Mayor

November 1, 2007

TO WHOM IT MAY CONCERN:

It is my pleasure to introduce internationally-known artist Mark Dion and his project, *William Bartram's Travels – Reconsidered.* Mr. Dion has recently been artist-in-residence at the Natural History Museum in London, and his most recent exhibit there *Systema metropolis* which opened last June to rave reviews.

For this project, he will examine the history and culture of 18th century American naturalists, John Bartram (1699-1777) and his son William (1739-1823). Using their journals, maps and drawings, Mr. Dion plans to retrace the exploratory journeys of the Bartrams, in particular, William's expedition to northern Florida. The artist will travel in much the same way as the Bartrams: by horseback, boat, and on foot. He will travel with other "explorers" to collect specimens and man-made artifacts found in this radically altered landscape and, in keeping with his longstanding interest in mail art, send them back to Bartram's Garden.

The connections between the Bartrams and Charleston are well-known. John Bartram visited several times between 1760 and 1765 (as the "King's Botanist") with his son William. William Bartram spent much time in the area of Charleston between 1766-77, 1772-73, and 1773-76. They both stayed with Thomas and Elizabeth Lamboll, whose house still stands in Charleston and is currently a bed and breakfast.

Of Charleston, John Bartram wrote, "*we were accommodated in the most civil and best manner*" William Bartram wrote to Mary Lamboll Thomas in 1786 that, "*a little Prayer-Book which you gave me When I departed from Your house, on the long Journey; & which, I always revered as the representative of that guardian sperit & protectress that attended me on those dangerous & dubious scenes in my Travels…*"

In the long-standing tradition of hospitality shown by the residents of Charlestown, I am certain that Mark Dion will receive the same *civil* accommodations and generosity of *sperit* as William and John Bartram over 200 years ago. I look forward to seeing the results of his explorations and vision in *William Bartram's Travels – Reconsidered.*

Most sincerely yours,

Joseph P. Riley, Jr.
Mayor, City of Charleston

JPR,jr./dm

CHARLESTON
All America City

P.O. Box 652, Charleston, South Carolina 29402
843-577-6970 Fax 843-720-3827

Gregory Volk
264 East 7th Street, Apt. 3
New York, NY 10009

New York City

The 9th of November, 2007

Dear Sir or Madam,

I beg to convey that the bearer of this document, Mark Dion, is an artist of remarkable achievement and impeccable repute. He is possessed of excellent manual skills, honed through years of education and diligent practice, and he is also possessed of a keen and inquiring intellect, which he brings to bear on his sculptures. Several estimable fields of knowledge, including Philosophy, Religion, Economics, History (both ancient and modern), and various branches of the Natural Sciences are all of utmost importance to Mr. Dion, and as much as his acclaimed sculptures are works of artistic excellence, they are also shining examples of scholarship and intellectual rigor. Moreover, Mr. Dion has a strong moral foundation and an excellent character grounded in timeless human values, such as generosity, fairness, kindness, forbearance, modesty, and respect.

Mr. Dion is now embarking on a noble excursion which will be arduous, but which he approaches with joy, optimism, and faith. I liken him to a bold explorer of yore, leaving the safety and familiarity of hearth and home to venture forth into the wide and uncertain world on an insatiable quest for pure knowledge. I humbly entreat you to assist him when possible, and to offer him helpful measures of comfort and support. You will, Sir, or Madam, discover him to be a delightful companion, an inspiring conversationalist, and, ultimately, an artist who, while familiar with fame and attention, is burdened by neither, but instead proceeds on his way with magnanimity, good humor, and a clear sense of purpose, born of understanding how uplifting and beneficial the plastic arts can be to us as individuals, as well as to our whole society.

I wish to thank you in advance for what generosity, encouragement, and support you may most graciously bestow on Mark Dion during his worthy excursion. I also believe, and have reason to believe, that this excursion will result in a work of art that proves inspirational for our era, and that will be beloved by generations to come.

Your humble and devoted servant,

Gregory Volk

November 9, 2007

To All Persons to Whom This May be Shown:

Greetings:

Mark Dion, the bearer is an artist, following in the trail of William Bartram. He is an honest and worthy man, traveling in search of what is uncommon, useful, and curious. I recommend him to your civilities. Receive him kindly and give him advice, and assist him with direction in travelling from one place to another.

In turn, Mark Dion's collections will be forwarded to Bartram's Garden in Philadelphia, organized, and exhibited for the benefit of the public. Aiding and assisting him is becoming to all encouragers of useful discoveries.

Sincerely,

Joel T. Fry
Curator
Bartram's Garden

JOHN BARTRAM ASSOCIATION
Bartram's Garden, 54th St. & Lindbergh Blvd., Philadelphia, Pennsylvania 19143
Phone: (215) 729-5281 · Fax: (215) 729-1047 · www.bartramsgarden.org

To the honourable soul it may concern,

May this missive find its beholder sound of shank and in good kidney.

The bearer of this letter is my charge, Mr. Mark Dion, an acolyte of the natural philosophies. Divine Providence has seen fit that I serve as Mr. Dion's instrument of ingratiation to your kind hospitality and generous protection. I would pray you might honor him with an open countenance and other such marks of Civility, Honour, and Regard, as Mr. Dion is a gentleman of eminence and esteem in his chosen profession, associations, and public employments — a man whose character is as unsullied as his shirtsleeves whilst supping at table. Hence, he is thusly recommended by his worthy patron to appeal to your wise counsel and assistance, for carrying into effect his intended travels, which shall take him far afield into the natural splendours of your fair lands, for the living things dwelling there are his object of study.

Any kindness, however small, shown to a man of such quality and promise as Mr. Dion shall not fall upon stony ground — but on the rich and abiding loam of Gratitude and Felicitous Reciprocity!

With every sentiment of deference,

Lord Brambrace Swells-ffWhimsy

Enthusiast and Gentleman Amateur

Departure Address

Bartram's Garden, Philadelphia

November 15, 2007

Ladies and gentlemen, friends, family, people of Philadelphia, fellow Pennsylvanians and those of you who have merely come to gawk, I come before you to bid you farewell, for my departure is imminent. Therefore I shall be brief in my remarks as the clock ticks away and the sun shall not dally for us and we intend to be South of Richmond on this day.

It has been deemed by the fine scholars of this esteemed institution, and by the wisdom of the holders of Philadelphia's cultural purse-strings and by the cautious connoisseurship of curator Julie Courtney – that I travel to the Southern territories of these United States to procure knowledse. This journey taken in seural stases and emparked upon by my faithful Campanion – Dana Sherwood, will undoubtably be one of hardship and great peril, however, we undertake this commission with utter determination of success. What, you may ask, will this success look like? What will it mean? While it is impossible for me to predict with percision the outcome of my endeavor, I am certain that I shall articulate a sensible understanding of the Southern landscape. For if we Americans possess Mastery of an artistic genere it surely must be the travelouge. While we can appreciate the literary importance of this genere in works as wide ranging as the accounts of Lewis and Clark to those of Jack Kerouac and Hunter S. Thompson, it is not clear what this genere means for sculpture. While painters like Fredric Church and Albert Bierstadt vividly respond to the travelouge, as do our most esteemed photographers like Danny Lyons, Ansil Adams and Walker Evans, but what about sculpture? In film, the road movie represents this form in the Bob Hope and Bing Crosby films, Easy Rider, Peewee's Big Adventure and the recent films of Gus Van Sants. Yet we are still at a loss as how to represent the transformative experience of travel through the material form of expression that is sculpture. It is my intent to vigorously interogate this question through braving the wilds of North Florida and west to the Mississippi River, and other areas unknown to myself.

I shall shadow the travels of fellow artist William Bartram. Bartram was a man of uncommon courage and a talent and intellect of formidable skill, particularly with regard to writing and draftsmanship. While in both categories I lack Bartram's eloquence, we do share the artist's vision. Therefore I pledge to do my very best to select, collect, preserve and artfully arrange the numerous natural objects and artifacts of material culture I encounter on this journey, with the purpose of display here in the house of Bartram for visual pleasure and intellectual edification of the people of Philadelphia. This charge I take on with awesome gravity. I shall explore regions remote and unknown to myself returning to this very nobel institution to report upon the wonderful nature of the region, as well as ~~the~~ on the curious ways, bizzare rituals and marvelous material culture of the humanity I encounter.

Deeply do I thank those who have assisted in my commission and I truely hope to do as little wrong as possible in the fullfilment of my duty.

Mark Dion addressing well-wishers at his send-off, November 2007.

Leaves gathered from Bartram's Garden on the day of departure.

From
Mark Dion
Travels of William Bartram
Best Western Hotel 337 East Beach Blvd
Gulfshores AL 36542
www.usps.com
FIRST CLASS
TO: M. Dion
Bartram's Garden
54th st. & Lindbergh Blvd.
Philadelphia, PA 19143
DO NOT O

Curious Cabinets

Julie Courtney

As an independent curator, my practice revolves around matching artists with sites. The sites are usually historical, sometimes abandoned or forgotten, often off the beaten track. Ideally it comes as a *Eureka!*—the perfect artist for the perfect site. Usually I have to track the artist down, see if there is interest, and then convince the site to allow us to create whatever intervention we have envisioned. However, one hot summer day in 2006 when I was visiting his home in northeastern Pennsylvania, Mark Dion said to me, "Why don't we do a project at Bartram's Garden?" I knew that instant it was the *Eureka!* moment I always strive for. Mark Dion, with his insatiable curiosity for the natural and the odd, and his ability to put things together to make the most astonishing *Wunderkammer*, was just the artist to work at this particular site. A project like this is just what I love to do: a perfect marriage between an amazing historic site in Philadelphia (although under-recognized, in the art world at least) and an extremely well-recognized contemporary artist. I knew that Mark would bring fresh insights and new attention to Bartram's Garden.

When I enthusiastically approached then-director Bill LeFevre with our idea, it met with approval as he recognized that this was just the kind of thing to bring Bartram's Garden into the forefront as a site for contemporary art. They had hosted contemporary exhibitions and performances sporadically, but this was the first time an internationally recognized artist would be working at the garden. A site like Bartram's Garden is ripe with possibilities. Their curator Joel T. Fry is a Bartram scholar and historian, while my background is contemporary art and working with artists. Bartram's Garden agreed to be the institutional sponsor and I would do the work of designing the program, finding collaborating institutions, and organizing and writing a proposal to the Philadelphia Exhibitions Initiative (PEI), who previously had underwritten my projects at sites like Eastern State Penitentiary.

Beginning with Mark Dion's handwritten letter of interest, in which he described the project as "part Lewis and Clark, part Jack Kerouac, part *Pee Wee's Big Adventure* and part *Borat*," I could see that many potential partners would appreciate the wit and humor the adventure promised. The premise was, more or less, to retrace travels by John Bartram (1699–1777), often considered America's first botanist, and his son William Bartram (1739–1823), especially William's famous journey from the Carolinas to Northern Florida. Mark was particularly drawn to William because he was a wonderful artist and his travels were well documented in his book *The Travels of William Bartram*. Like Bartram's, Mark Dion's journals, too, are documents of keen observation and drawings of what is seen and experienced. Of course the landscape of 21st century is vastly different from the landscape Bartram traveled, but that would be part of the surprise, adventure, and irony of Mark's trip.

John and William Bartram traveled by boat, on horseback, and on foot. Mark Dion basically traveled by car. Since he does not drive, this task was ably carried out by his traveling companion and fiancée, Dana Sherwood. The Bartrams collected and shipped back specimens of plants and animals. William often painted what he saw, and drawings of his are in numerous museum and library collections. Mark would also collect plant and animal specimens and water samples, and, like the Bartrams, would press hundreds of plants. And he would collect thousands of objects, including examples of 21st-century material culture, for display.

An important part of the exhibition would be the cabinets that would house the collections: the Cabinets of Curiosity Mark is so well known for. He would use cupboards and cabinets already in the house, would buy similar old ones, and three would have to be built. Mark made drawings for a large multidrawer cabinet, a postcard cabinet, and a replica of the cabinet for housing botanical specimens designed by Carl Linnaeus (1707–1778), the Swedish botanist and contemporary of John Bartram who was the father of binomial nomenclature. The drawings are not typical 21st-century computer-generated "shop" drawings, but were done in Mark's typical hand-made fashion with blue and red colored pencils. Though they were not what the

Tools of the artist explorer include herbariums, dissection kits, entomological pins and drying boards, nets, hats, zip-lock bags, knives, pencils and pens, tools for measurement, cameras, scanner, computer, sleeping bag, tent, propane stove, fishing poles, et-cetera.

Scrounging for objects of interest in a flea market outside Charleston.

cabinetmakers were used to, the drawings had the dimensions and all information required.

When we were awarded the exhibition grant from PEI, we were also told that we would receive additional support from a new Marketing Innovation Program. With this additional boost we would be able to have a website that would track Mark's travels. Photographs, videos, and various images could be put up on a site while he was on the road. For an admitted Luddite like Mark, this was a total antithesis to the way he works and presented a big challenge. As Mark says, "I like *things*—not pictures or facsimiles of things." Thank goodness for Dana. Equipped with a new camera, GPS device (which, as I witnessed, Mark nearly threw out the window of their New York City apartment), scanner, and video camera, Dana—who dresses like a melancholic 19th-century poet but has the knowledge of a 21st-century technologically savvy adventurer—would be able to record and document their travels.

We knew that *Travels of William Bartram—Reconsidered* could be an accessible and engaging project for connecting with the community. Richard Allen Preparatory Charter School is just down the street from Bartram's Garden, and Bartram's staff members Melanie Snyder, Stephanie Phillips, and I went to tell a group of middle schoolers about William Bartram and Mark Dion. We had 45 minutes to tell their stories and help the students begin to create art projects. Using beautiful wooden cigar boxes donated by Holts Cigars, art supplies, and other things that might be useful to an explorer of Mark's experience, students threw themselves into making boxes that they could present as going-away presents.

On a rainy November day in 2007, many well-wishers gathered to send Mark on his way. It was a brilliant beginning to the travels. Mark and his intrepid traveling companion and technology expert had laid out all of the equipment they planned to use on their travels. Camping gear, art-making supplies, fishing and butterfly-catching equipment, bug-collecting materials, flower presses, his and hers swamp boots and hats, cooking equipment and the requisite single malt Scotch were but a few items that covered a table that extended the width of the carriage house at Bartram's Garden. It was a feast for the eyes, and a wonderful way to introduce what we might expect from an artist like Mark Dion. Lord Whimsy, local dandy and carnivorous-plant expert, presided, and after reading from his book *The Affected Provincial's Companion, Volume I,* he read letters of introduction prepared for Mark by such luminaries as Mayor Joseph P. Riley Jr., of Charleston, SC; Sue Ann Prince, museum director at the American Philosophical Society (founded by Benjamin Franklin along with his very good friend John Bartram), John Bartram Association board member Jim Straw; and the art critic Gregory Volk. The students from Richard Allen Preparatory Charter School had walked down to the garden in the pouring rain, and several girls presented the gifts to Mark. He was visibly moved by the creativity of the work.

An example of the postcards Dion sent back to Philadelphia.

After the travelers set out, packages and hand-painted postcards began arriving. The postcards had images of such things as pottery shards, birds, critter-eaten leaves, Mickey Mouse, and golf carts, with brief descriptions. These were addressed to the Garden's staff and various people connected to the project. We were instructed not to open any of the packages, but anticipation was great and we all looked forward to the travelers' return.

The website also began to come alive. Stories, photographs, maps and videos began appearing and the site became a parallel project. The project was incredibly well documented—thanks to Dana—on the site www.markdionsbartramstravels.com, and we could follow his whereabouts and read about his adventures, recounted in much the same way William documented his travels, though to a much larger audience and during the actual trip rather than years after.

Traveling from Philadelphia through Charleston and beyond, Mark and Dana collected plant and animal matter and lots of material culture that he found and purchased in flea markets and yard sales. He is a champion shopper with a sharp eye. They returned from the first leg of their travels in January and came to Bartram's Garden to open boxes and see what they had and how much more they would need to collect. Boxes with water samples made it through the watchful eye of the postal service, whose first question is always, "Is there anything liquid in these boxes?"

Although the clerks were puzzled, they shipped them, and the numerous plastic bottles arrived intact. Boxes stamped with "Hate Archives" were placed on a shelf and are never to be opened. The Bartrams themselves ran into many alligators, and Mark collected numerous specimens of souvenir alligators, enough to fill a nearly wall-sized glass case. The range of these, from ashtrays to shot glasses to toys to nutcrackers is astonishing and delightful. When looking at William Bartram's drawing and written description of the alligator he encountered, it is obvious that he did not spend a lot of time looking at or sketching the beast.

Mark's second journey in spring of 2008 saw his marriage to Dana and a residency at the Atlantic Center for the Arts in New Smyrna Beach, Florida, the site of William Bartram's shipwreck, which is only loosely referred to in *Travels*. There Mark worked with eight "associates"—artists from the US, Quebec, and Europe, who had been selected from a group of 200 applicants. All of these artists became part of his team, and soon dead birds, reptiles, exotic roots, and plants were assembled and added to the collection for the exhibition. Many exploratory journeys, including one to Disney's Animal Kingdom Theme Park—which I, for one, hadn't contemplated but was fortunate enough to be part of—added to our enlightenment. I have to admit, we did see some extraordinary bats—I'm not sure they were native to the area or that the Bartrams would have seen such astonishing creatures.

As did William Bartram, Mark documented everything in his journal—drawings, lists of birds spotted, accounts of what was seen, heard, and tasted. Mark's book also contains receipts and photocopies of restaurant cards, editorial comments on and drawings of meals consumed, specific dates and times and names of fellow diners. In the journal are explicit route numbers and drawings of things ordinary and extraordinary, with detailed accounts of the activities of any given day. Everything, no matter how mundane (e.g., small hotel soaps and creams) could become part of the installation, and be installed with the same reverence as a rare bird.

The spectacular journey came to an end, and the real work of cleaning and assembling began. Mark's assistants Valerie Piraino and Mia Helmer, along with Dana, Willa Courtney, and I, put everything in place and got the house in shape for the opening. The magic of Mark Dion's art culminates in arrangements and unlikely juxtapositions, in this case from the thousands of objects collected on his journey. Drawers are filled with objects both remarkable and unremarkable—beautifully preserved birds, shells, nuts, and pinecones are placed next to rusty nails, corks (from all the wine consumed on the travels), small toys, magic lantern slides, bottlecaps, and things too numerous to mention. The sheer quantity is enough to overwhelm, but Mark's talent for arranging the findings in the most elegant way reveals his gifts and inspires awe in viewers. Inserting a 21st-century cabinet of wonder in an historic house, Mark Dion helps us see what might be ignored and guides us to reexamine our past.

A project like this can never happen without serious financial support, and my thanks on behalf of all involved go to Paula Marincola, director of the Philadelphia Exhibitions Initiative and now of the Pew Center for Arts and Humanities (PCAH) for her faith in this project. Paula's forward thinking allows projects like this to flourish and encourages curators and institutions to break the boundaries. Roy Wilbur, of the Marketing Innovation Program of the Philadelphia Cultural Management Initiative, also a program of PCAH, provided marketing support and good ideas. There is never enough for outreach, and this support provided the website and publicity which helped us reach thousands of people from all over.

I want to thank Bartram's Garden and the staff there for their generosity in allowing us in the historic house built by John Bartram. Louise Turan, the new executive director, inherited the project, and recognized its implications for the future. Curator Joel Fry was particularly supportive, I think because he saw in Mark a serious fan of William Bartram who would treat the house and history with respect. Not many historians would allow such freedom. He rarely cringed—except upon the unleashing of the voodoo curse of a gris-gris I accidentally spilled. And thanks to all the others of the extremely competent staff members for extending their already full work load. From the moment we got the funding, Melanie Snyder, director of education and public programs, and Stephanie Phillips, director of development, were always available and good-natured about whatever came up, and I don't know what I would have done without them.

Some seeds collected during the first stage of the expedition.

November 19, 2007

On Monday November 19th we were afforded an opportunity to observe a tribe of people who live in a manner utterly foreign to us. They live in a village fortified with fences and moats called Sun City. The name is appropriate since these people seem particularly fixated on the topic of weather and greatly speak with disparagement regarding the climates of their native birthplace in the North. Indeed, these migrating folk mostly originate from Ohio, Massachusetts, New York and Pennsylvania. They are by the majority fair in complexion,

(continued on page 17)

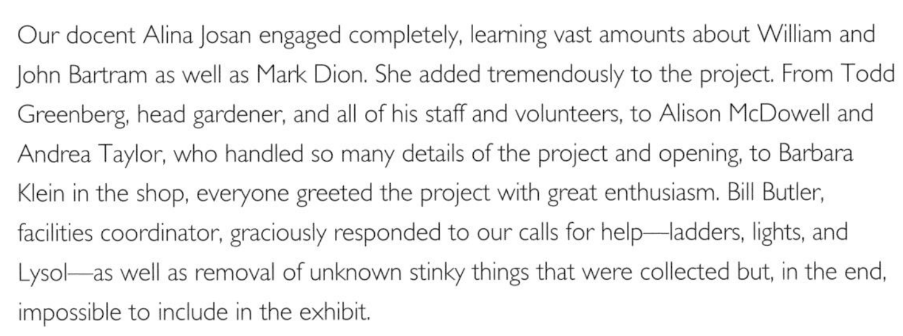

Our docent Alina Josan engaged completely, learning vast amounts about William and John Bartram as well as Mark Dion. She added tremendously to the project. From Todd Greenberg, head gardener, and all of his staff and volunteers, to Alison McDowell and Andrea Taylor, who handled so many details of the project and opening, to Barbara Klein in the shop, everyone greeted the project with great enthusiasm. Bill Butler, facilities coordinator, graciously responded to our calls for help—ladders, lights, and Lysol—as well as removal of unknown stinky things that were collected but, in the end, impossible to include in the exhibit.

We engaged Ian and Matt Pappajohn and Kristine Kennedy of Pappajohn Woodworking to make the three new cabinets. Their ability to interpret Mark's rather sketchy drawings (which nevertheless had very explicit dimensions) to create such gorgeous cabinets was impressive.

During the exhibition period three collaborative events were staged with Mark and other Philadelphia institutions, and we would like to thank Susan Glassman of the Wagner Free Institute of Science, Kim Sajet of the Historical Society of Pennsylvania, and Roy Goodman and Sue Ann Prince of the American Philosophical Society. Mark's and Dana's three trips were facilitated by the hospitality of a number of friends and temporary traveling companions. The two of them would like to extend their gratitude to: Hope Ginsburg, Tom Sherwood, Susanna Faver, Raven Hinojosa, Jayme Kalal, Courtney Lain, Pandora Gastelum, Nina Nichols, and all their numerous comrades in New Orleans who made the travelers' time there unforgettable. Thanks to those at the Atlantic Center for the Arts: director Ann Brady, Nick Conroy, Jim Frost, and the rest of the dynamic staff as well as to Mark's fellow seminar leaders, David Lang and Susan Marshall. The group associate artists Corrine May Botz, Barbara Breitenfellner, Bronwen Buckeridge, Scott Hocking, Vincent LaFrance, Nancy Anne Mcphee, Christopher Patch, and Filip Van Dingenen contributed many fine treasures in the installation. Gratitude is also extended to the rest of the Florida support team: Nitin Jayaswal, Neil and Janna Sherwood, Graphic Studio of Tampa, and Dana's grandma, Adele Sherwood. In addition, thanks to the fine sportsmen of Alabama, who rescued Mark, Dana, and their reliable and even-tempered companion Christy Gast on more than one occasion.

I am most grateful to Dana Sherwood, Mark's right-hand woman and his guide into the 21st century. In addition to not typing, which I deduced when I received his handwritten letter of interest, Mark does not use a computer, so you can imagine, with a project like this, where I promised daily blogs in our grant proposal, how absolutely essential Dana was to the project's success. She provided the fabulous photographs and videos posted online, typed Mark's journal entries for the website, and provided a level head, warmth, and grace.

The incredible website was created by the highly creative team of Laris Kreslins and Kendra Gaeta (with assistance of Tom Bubul, Alexis Lerro, and Lord Whimsy) of Lime Projects, whom Roy Wilbur introduced us to. They taught me a new language and created a thoughtful and entertaining site that will live long after the exhibition comes down. Similarly, Braithwaite Communications, a public relations firm, handled the send-off as well as the opening. Cass Oryl was incredibly responsive to our requests as well as those of the press, and we are thankful to her for doing such a great job.

Jeffrey Jenkins's creative design of all printed materials are lasting emblems of *Travels of William Bartram—Reconsidered* and I am in awe of his generosity and responsiveness to our needs. Joseph Newland, initially another gift from the Exhibitions Initiative as a consultant, oversaw the organization and editing of the catalogue, and I am particularly grateful to him for his advice. Aaron Igler's amazing photography of every little tiny object communicated the enormity of this undertaking, and I can't imagine that there were enough hours in the day to accomplish what he did. Gregory Volk's insightful essay is the verbal version of a cabinet of curiosity, and has made this book a truly memorable document.

Lastly, and of course firstly, I am indebted to Mark. A brilliant artist with a subtle sense of humor and irony, Mark has taught me how to look at life in a different way. Getting him to find time so often in the last year hasn't been easy. He traveled around the world—this year alone he was out of the country more than he was in it—in addition to his undertaking the *Travels of William Bartram—Reconsidered*. Beyond the trips he took for the project, the material he had to provide for the website added an enormous amount of work that none of us had anticipated, seriously cutting into his time for flea marketing and archeological exploration. As with most things, he took this on with grace and elegance. He is articulate, thoughtful, and a joy to work with. Something of a magician, Mark even had me cozying up to a snake in a jar.

The houses are uniform in color, each resembling the beige one might encounter used in a hotel telephone.

(continued from page 15)

though often red from overexposure to the sun. While healthy, it is curious to note that no children or youths were encountered on our excursion to this land. Our visit commenced at a security stockade where a brisk elder took great pleasure in instructing us to take our first right and fourteenth left. Through curvy main roads intersected by feeder streets named after Robert Lee and his military colleagues, we slowly wander our way past artificial lakes, patches of woodlands and houses. All this time we were conscious of our speed due to the numerous video surveillance cameras.

The dwellings of the locals are spacious and of recent construction. In fact, no structure in this settlement seemed old or evidenced decay or use. The habitations seemed to be the architectural plan of a single person. The houses are uniform in color, each resembling the beige one might encounter used in a hotel telephone. Before each dwelling was a conspicuously brown, neat lawn planted with manicured shrubs and tropical foliage. Apparently, despite the fountains, ponds and other waterworks, the southern drought has had some effect here on the landscape.

"Two rituals bind the people here to this place. They are the games of golf and tennis, which have been elevated to the status of a religion. The game of golf requires considerable amounts of land and significant engineering works, while the tennis courts are built into a single compound. These activities largely construct the time of the inhabitants, however, there are numerous other distractions from swimming to the social drinking of alcohol. There is even a nature path which curiously is a walkway elevated off the forest floor. Perhaps the intention for this is to keep those who would appreciate nature clean since hygiene seems to be of the utmost importance to these singular people. Indeed numerous signs of the village warn that pet droppings spread disease.

"These people are active, although seem to work little. Indeed labor such as maintenance, garden tending and cleaning is not performed by the denizens of the village, but by workers imported from beyond the compound. Many of these laborers are much younger and are often ethnically distinct from the Sun City villagers, who on the whole, are a cultural, ethnic and generational monoculture.

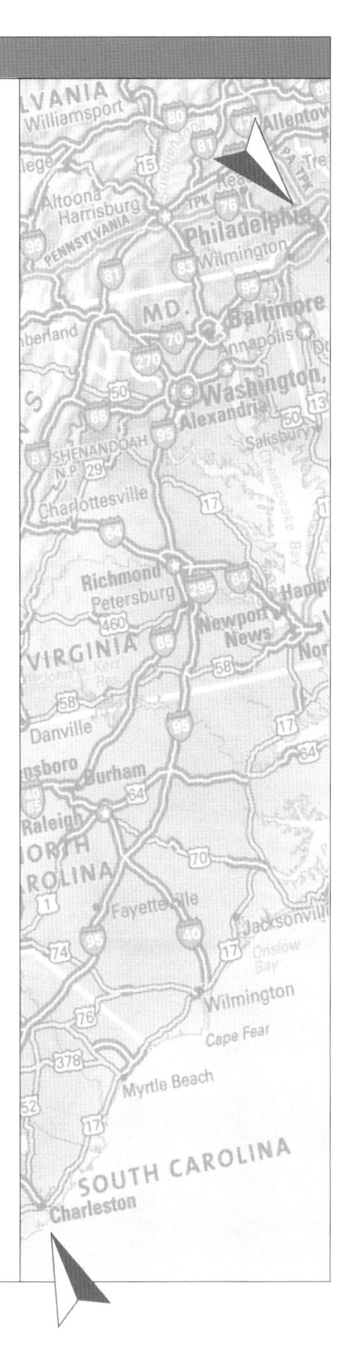

This is a very ancient garden. . .

— Reverend Manasseh Cutler, 1787

A Short History of Bartram's Garden

Joel T. Fry

Bartram's Garden was founded in the autumn of 1728 when John Bartram (1699-1777) purchased an improved farm in rural Kingsessing Township, part of Philadelphia County. Bartram, a third-generation Pennsylvania Quaker from nearby Darby, acquired a 102-acre farm with a "tenement," and two additional small pieces of marsh land, at a sheriff's sale on September 13, 1728, for £145.[1] The farm had been part of a much larger plantation on the west bank of the lower Schuylkill once known as Aronameck, first occupied by Europeans in 1648 during Swedish colonial settlement of the Delaware Valley.

John Bartram suffered a series of tragedies in his early life. He lost both of his parents at a young age. His first marriage was not fated to last long, as his first wife Mary Maris died in 1727, leaving John a widower with one surviving son, Isaac. A year after her death, Bartram purchased the new farm in rural Kingsessing Township, and one year later, on December 11, 1729, he married Ann Mendenhall (1703-1789) at Concord Meeting in Chester County. The new couple soon moved to the Schuylkill River farm.[2] John Bartram constructed a new stone house and auspiciously marked the work with a date stone, **"IOHN❖ANN:BARTRAM:1731"** with the inscription **"GOD SAVE"** in Greek characters above. John and Ann Bartram established what came to be a dynasty of American naturalists and gardeners; four generations of the Bartram family would live and work at what became the Bartram botanic garden.

It seems likely that Bartram chose this favorable site with the intention of establishing a large garden, and the location remains well suited to the cultivation of plants today. The initial garden was probably laid out at six or seven acres, on a terraced slope, leading from the house to the river. It was expanded under succeeding generations. Additional space was set aside for an orchard, greenhouses, garden framing, and nursery beds, and the complete garden may have totaled as much as twelve acres at its peak in the 1830s. John Bartram's garden began as a personal garden, but grew to a systematic collection of native and exotic plants as he devoted more time to exploration and discovery. Exchanges of plants and seeds from gardens in North America and abroad also fueled the collection. Although not the first botanic collection in North America, by the middle of the eighteenth century, Bartram's garden contained the most varied collection of North American plants in the world.

About 1733, in an event important to the history of horticulture and natural science, John Bartram began a correspondence with the London cloth merchant Peter Collinson (1694–1768). Collinson, a member of the Royal Society—and like Bartram a Quaker and an enthusiastic gardener—became the middleman to a scientific trade in seeds, plants, and natural history specimens.[3] Plants from Bartram's garden were exchanged with a range of botanists, gardeners, and nurserymen in London, and throughout Europe. Collinson also arranged funding from patrons among the British elite. During his career John Bartram traveled widely throughout the British colonies in North America—his plant collecting began in the Mid-Atlantic colonies of Pennsylvania, New Jersey, Delaware, and Maryland. In time, Bartram traveled north to New York and New England, and south to Virginia, the Carolinas, Georgia, and Florida, exploring a region that spanned from Lake Ontario in New York to the St. John's River in Florida, and from the Atlantic coast to the Ohio Valley.

As John Bartram tended his garden, he established a family institution that survived him and grew under the care of three generations of his family. Following the American Revolution, Bartram's sons John Bartram Jr. (1743–1812) and William Bartram (1739–1823) continued the international plant trade their father had established, and expanded the family botanic garden and nursery business. Following in his father's footsteps, William Bartram became an important naturalist, artist, and author in his own right, and traveled the American South from 1773 to 1776 under the patronage of Dr. John

A Catalogue of Trees, Shrubs, and Herbaceous Plants, Indigenous to the United States of America…
Philadelphia: 1807

left, "*A Draught of John Bartram's House and Garden as it appears from the River.*" *Ink and wash drawing by William Bartram, 1758.*
above, *East porch and "Common Flower Garden," Bartram's Garden.*

William Bartram, (1739-1823). Portrait in oil by Charles Willson Peale, 1808.

Fothergill. William Bartram's *Travels...* published in Philadelphia in 1791, and reissued in a number of European editions, strengthened the connection between the name Bartram and the science of plants in North America.[4] Under William Bartram the garden became an educational center and helped to train a new generation of natural scientists, artists, and explorers. In the early Federal period the Bartram Botanic Garden served as *the* American botanic garden in lieu of any official institution in Philadelphia

From 1809 onward, Ann Bartram Carr (1779-1858), a daughter of John Bartram Jr., continued the family garden. Ann B. Carr had been educated by her uncle William and inherited his skill for illustration as well as the family passion for plants. With her husband, Colonel Robert Carr (1778-1866), her stepson John Bartram Carr (1804-1839), and a nephew, George Gale (born ca. 1816), the international trade in seeds and plants was continued. The garden was considerably enlarged as a commercial nursery, and at its peak featured ten greenhouses and a collection of over 1400 native plant species, and as many as 1000 species of exotics, many under glass.

Financial difficulties led to the sale of the family garden by the last of the Bartram descendants, in 1850. The new owner, Andrew M. Eastwick (1811–1879), a wealthy railroad industrialist, preserved the historic garden as a private park on his estate. At Eastwick's death in 1879, the expansion of the city of Philadelphia and the movement of industry to the lower Schuylkill threatened the garden site. A campaign to preserve it was organized by nurseryman and writer Thomas Meehan, in Philadelphia, and Charles S. Sargent, of the Arnold Arboretum in Boston. In 1888, after a long political fight, the historic garden was placed on the city plan and slated for preservation. In 1891 the city took possession, and the garden has been protected as a city park since then.

The First Generation: John Bartram, His Garden, and His Business

John Bartram was imbued with the Quaker curiosity and reverence for nature. There is much evidence that Bartram's interest in botany and natural history began at a young age. Bartram wrote his friend Collinson, late in their correspondence:

> I had always since 10 years ould A great inclination to plants & knowed all that I once observed by sight tho not thair proper names haveing no person or books to instruct me.[5]

And Bartram wrote Sir Hans Sloane that "Physic and Surgery...was my chief study in my youthful years."[6]

James Logan of Philadelphia wrote Collinson in 1736 that "Bartram has a genius perfectly well turned for botany and the Productions of Nature." [7] And in the summer of 1737 he repeated John Bartram had been "formed a Botanist by nature which I never was knows the kind and name of every plant he sees or at least most that have occurred to him."[8]

It is perhaps not surprising that Bartram gravitated to the circle of Benjamin Franklin and his curious friends in the city of Philadelphia. Bartram was at some remove from the colonial city, and so it's not likely he attended the weekly gatherings of Franklin's Junto. But in the early 1730s Bartram became a good friend of the "good-natur'd" Joseph Breintnall (d. 1746), who was one of Franklin's most active collaborators. It was Breintnall—merchant, copyist, poet, and first secretary to the Library Company—who recommended Bartram to Peter Collinson.

Benjamin Franklin also probably came to know Bartram through Breintnall, and Franklin was soon publicizing him. He reported Bartram's discovery of ginseng on the Susquehanna in 1738, and published useful articles on plants by Bartram in his newspapers and almanacs. Franklin wrote his parents in Boston, "We have a botanist here, an intimate Friend of mine, who knows all the plants in the Country." [9] In the spring of 1742 Franklin proposed a colonial subscription to fund Bartram's travels.

> John Bartram has had a Propensity to Botanicks from his Infancy, and to the Productions of Nature in general, and is an accurate Observator, well known in Pennsylvania, where he was born and resides, to be a Person fitted for this Employment.[10]

And in 1743 Bartram and Franklin together founded the American Philosophical Society.[11]

It has often been repeated that Linnaeus dubbed Bartram "the greatest natural botanist in the world."[12] Unfortunately, this quotation survives without any original context, but at face value, it is a surprisingly accurate classification for the Quaker botanist. Pehr Kalm (1716-1779), a Swedish student of Linnaeus, visited Bartram on a

opposite page, 1- *Detail: Nicholas Scull and George Heap, "A Map of Philadelphia and Parts Adjacent...," engraved by Lawrence Herbert, Philadelphia: 1752.*
2- *Advertisement, Philadelphia Aurora, June 27, 1811.* 3- *Window inscription, Bartram house. Souvenir postcard, ca. 1907.*
4- *Signed flyleaf, "W. Bartram His Book June the 10 1755- " Linnaeus, Genera Plantarum, 2nd ed., Leiden: 1742.*

BARTRAM'S GARDENS, KINGSESS.

FOR SALE, A VARIETY OF

GREEN HOUSE PLANTS,

AMONG WHICH ARE

GREEN and BOHEA TEA, Cape JESSAMINE, DAPHNE INDICUM, HYDRANGIA HORTENSIS, HYDRANGIA QUERCIFOLIA, PERPETUAL FLOWERING (*Otaheitian*) ROSE, VERBENA TRIPHYLLA, NUTMEG GERANIUM, JUSTICIA, &c. &c. all in a thriving condition.

JOHN BARTRAM

HEMP.

A FARMER, or other person, accustomed to the culture and water-roting of HEMP, may hear of a very advantageous situation, by appling at the above Gardens, or at R. & W. Carr's Printing Office, No. 51, Sansom street.

June 27 ifws2w

Peter Collinson (1694–1768) (Engraved portrait of Peter Collinson from "Some Account of the Late Peter Collinson," London: 1770.)

Collinson, a member of the Royal Society–and like Bartram a Quaker and an enthusiastic gardener—became the middleman to a scientific trade in seeds, plants, and natural history specimens.

number of occasions between 1748 and 1751, during his stay in North America. Kalm was taken with Bartram's "peculiar genius" for natural philosophy and natural science.

> Mr. John Bartram, an Englishman, who lives in the country, about four miles from Philadelphia, has acquired a great knowledge of natural philosophy and history, and seems to be born with a peculiar genius for these sciences. In his youth he had no opportunity of going to school. But by his own diligence and indefatigable application he got, without instruction, so far in Latin, as to understand all Latin books, and even those which were filled with terms.[13]

Philadelphia was in an ideal location to found a North American botanic collection. The Mid-Atlantic climate was moderate and plants from far north or far south could survive in the open ground. The city was located along the boundary between two geographic regions—the coastal plain and piedmont. Bartram's Kingsessing farm sat squarely on this boundary, and was already host to a varied natural plant community. But within a short distance were diverse natural environments, each with characteristic plants. Starting locally, Bartram explored the New Jersey pinelands and coast, the sources of the Schuylkill River and mountains of Pennsylvania, the three Lower Counties of Delaware, and even the Eastern Shore and upper Chesapeake. These nearby collecting areas became the source of much of the stock for the yearly seed business, and he and his family returned to these locations often, particularly to southern New Jersey.

John Bartram's travels were partly funded by a business in North American seeds and plants organized through his friend and chief correspondent, Peter Collinson in London. With Collinson as the middleman, seeds and dried specimens as well as animal and mineral specimens and other curiosities were sent to Philip Miller at the Chelsea Physic Garden, Johann Jakob Dillenius at Oxford, Mark Catesby and Sir Hans Sloane in London, Johan Frederik Gronovius at Leiden in the Netherlands, and Linnaeus in Sweden.[14] During Bartram's life, European scientists were in the process of codifying a new system of scientific botany, largely based on the work of Linnaeus. Bartram was able to provide hard evidence on the plants of North America, and his contributions were mentioned by all these authors. In return Bartram received publications from Collinson and his friends in the natural science community, which formed the basis of a valuable reference library at the Bartram garden.

Peter Collinson's connections with the British horticultural world also led to lucrative commissions from gardens for North American seeds. In the 1730s this included a select few from the landscape garden movement in England: Robert James, eighth Baron, Lord Petre, at Thorndon Hall in Essex; his cousin, Edward Howard, ninth Duke of Norfolk, at Worksop Manor; and Charles Lennox, second Duke of Richmond, at Goodwood. By the middle of the 1740s North American plants became increasingly fashionable, and Bartram's customer list grew to include the owners of many well-known British gardens: John Russell, fourth Duke of Bedford, at Woburn; Archibald Campbell, third Duke of Argyll, at Whitton; John Stuart, third Earl of Bute; Norborne Berkeley, Lord Botetourt at Stoke Park; the ninth Earl of Lincoln at Oatlands Park; Charles Hamilton at Painshill Park; Dr. John Fothergill; and many others.[15] From the beginning of their friendship, Bartram made up special boxes of live plants and roots from his latest discoveries for Collinson alone. These plants quickly made Collinson's garden famous. Plants from Bartram were drawn by George Ehret and Mark Catesby, and published descriptions of these new North American species appeared in works by Catesby, Philip Miller, and Linnaeus, generally crediting Collinson.[16]

By the end of his career John Bartram had traveled widely throughout Pennsylvania and New Jersey, collecting in the Susquehanna Valley, the Pennsylvania Mountains, the Delaware Water Gap, and as far west as Pittsburgh and the Ohio River. Bartram traveled through Delaware and the Eastern Shore of Maryland; through the Virginia Tidewater, Piedmont, and Blue Ridge; in New York up the Hudson River Valley to

William Bartram, watercolor & ink on paper, early 1750s, "Black-throated Green Warbler (Dendroica virens), on Red Oak (Quercus rubra)".

Albany and the Catskills; and to Connecticut; Rhode Island; North Carolina; South Carolina; Georgia; and Florida. The journals of his trip from Pennsylvania to Onondaga in 1743, and of his investigations in Florida from St. Augustine to the source of the St. John's River in 1765–66, were published in his own lifetime.[17] Eventually through efforts of Collinson and also Franklin in London, John Bartram was awarded a yearly stipend of £50 from the king of England, George III, from 1765 to his death. Bartram was often styled "King's Botanist" in his later years.[18]

Only a single eighteenth-century illustration is known of the Bartram house and garden, entitled *A Draught of John Bartram's House and Garden as it appears from the River.*[19] This schematic view of garden in late 1758 is now generally thought to be by young William Bartram. It is an attempt at a plan and perspective drawing at same time, and so is not wholly true to scale, but it does record a wealth of detail. When looking at the natural landscape, the drawing shows the garden divided into two major parts—an upper garden on the terrace adjacent to the house and a larger lower garden running east to the river, below the terrace wall. The artist specified several areas: an "Upper Kitchen Garden," a "Common Flower Garden," a "New Flower Garden," and a "Lower Kitchen Garden." These plots are all fenced or defined by the terrace. The main collection of large specimen trees is planted along two long walks that run the length of the lower garden. A "Pond" in the center of the lower garden is feed by a "Spring head convaid underground to the Spring or Milk House." Although these plots are defined, there is little detail of the actual plantings. Stylized trees line the allées along the south edge of the lower garden, and trees are also indicated along the north fence, possibly espaliers. A few other isolated trees may mark important specimens, including a Bald Cypress, *Taxodium distichum*, at the southwest corner of the springhouse that survived as a landmark at the garden into the twentieth century. The overall impression of this landscape is that of a personal garden. To signify the owner of the garden, William drew his father standing in the lower garden surveying his domain.

"Old Florida Cypress Bartram's Garden," ca. 1875. *Photograph by John Moran. The figure beneath the tree may be Andrew M. Eastwick, who purchased the garden in 1850*

At some point in the 1750s John Bartram decided to enlarge his house and add a new classical façade. The 1758 drawing shows that the enlargement was already under way. Bartram was also disowned by his local Quaker meeting in 1758 for his deist or Unitarian views.[20] The fashionable new façade might be subtle symbol of his new religious independence. In any case, the new house front emerged over the course of the 1760s. In a letter to Collinson dated May 30, 1763, Bartram refers to a vine that "ran near 30 foot up a stone pillar last year."[21] This "pillar" is almost certainly one of the three stone columns which ornament the porch of the river façade of the house. Bartram probably completed his classical additions ca. 1770, and a second date stone and inscription on the east front of the house runs as follows:

IT IS GOD ALONE, ALMYTY LORD
THE HOLY ONE BY ME ADOR'D
IOHN BARTRAM 1770

John Bartram Sr. retired in the spring of 1771, although he had previously begun dividing his estate among his children. In late April 1771 he wrote Franklin: "My eyesight fails me very much and I am going to thro all my business into my Son John's hands except part of my garden."[22] Son John married a cousin, Elizah Howell, in May 1771. Following this marriage, John Sr. and his wife Ann moved into the southern half of the expanded Bartram house, and John Jr. and Elizah occupied the northern portion. The elder Bartram apparently lived quietly at his garden until his death on September 22, 1777, just before the British occupation of Philadelphia following the Battle of Brandywine.

Details: east facade of the Bartram house. The rustic Palladian decoration was carved by John Bartram in the 1760s.

Through the fall of 1779 the Bartram's packed several boxes for the French ambassador, Sieur Gérard. One of these boxes for Sieur Gérard included two of the original seedlings of the as yet unnamed Franklinia, "*to be planted in the Royal garden at Versailes*."

The Second Generation: John Bartram Jr. and William Bartram

William Bartram was arguably John Bartram's most famous son and a naturalist of renown in his own right, but he did not inherit the family garden and farm in Kingsessing. William had accompanied his father on a number of collecting expeditions, most notably the trip south to Florida in 1765. After a number of false starts in business and a disastrous failure as a planter in Florida,[23] William received the sponsorship of Dr. John Fothergill for exploration of the Carolinas, Georgia, and Florida. From the spring of 1773 through January 1777, William Bartram was on the longest and most important trip of discovery and natural research of his life.[24] After his return to Philadelphia, William spent the rest of his life at the garden. Several years were spent compiling an account of his travels; publication of this work was delayed a number of years, but it was eventually printed in Philadelphia in late 1791.[25] It has since become a classic in travel literature, in natural description, and for its early romantic literary style.[26] William led the life of a retired, Enlightenment sage at the garden. A compound fracture of a leg suffered as a result of a fall while gathering cypress seeds in the garden, ca. 1786, may have helped to enforce his retirement.[27]

The exact business relationship between William and his brother John is not known, but from surviving manuscripts it appears William tended to the written and academic side of the partnership, while John continued the annual gathering trips. Both worked cultivating the garden, and both probably packed seeds and plants for shipment abroad. Surviving plant lists, catalogues, plant descriptions, and bills from this period are all in William's hand. William almost certainly prepared the text for the printed catalogues issued by the garden between 1783 and 1819.[28] William also continued to describe and draw new plants for science. He prepared a number of illustrations for publication, notably for Benjamin Smith Barton's *Elements of Botany,* the first American text on the subject, published in 1803.

The American Revolution, and its associated European conflicts, disrupted international trade, and of course the trade in seeds and plants. The Bartrams were largely cut off from their European customers for at least eight years. In early 1779 a small trade with France opened, protected by French convoys. In March, John and William Bartram prepared a small box of twenty-two varieties of roots and seeds and two large boxes of over forty varieties of plant roots "for his Exelly the French Minister."[29] Through the fall of 1779 the Bartram's packed several boxes for the French ambassador, Sieur Gérard. One of these boxes included two of the original seedlings of the as-yet-unnamed Franklinia, "to be planted in the Royal garden at Versailes."[30]

There was a large demand for North American seeds and plants in France, particularly as the war with Britain had cut off usual sources of plant material. Franklin also suggested that the Bartrams "send the same number of boxes here, that you used to send to England, because England will then send here, for what it wants in that way."[31] With the end of American hostilities in 1781 and the Treaty of Paris in the spring of 1783 bringing European peace, John and William Bartram issued a catalogue. The Bartram *Catalogue* of 1783 was a simple, one-page printed list, a "broadside" of plants and seeds available for sale from their garden. Franklin arranged for republication of the list in Paris. This single large sheet is the first botanic list of North American plants to be printed in America, and is also one of the earliest known nursery catalogues from the United States. In botanic shorthand, the catalogue described the native plant collection that was unique to Bartram's Garden.[32]

There are few detailed records for the garden business, but a scattered series of manuscript catalogues suggests the Bartram brothers ran a successful business in the last two decades of the eighteenth century. The Bartrams were involved in the planting of Samuel Vaughan's State House Garden in the 1780s, an innovative public garden in Philadelphia.[33]

1783

CATALOG

S D'ARBUSTES ET

Amérique, & produiſent des Graines en Mat

de Philadephie; qui ſe vendent en Plantes

TRES DE RENVOI. a Un bon Terreau humide ſur la Glaiſe

& humide dans les Montagnes pierreuſes. d Un Terrein ſa

nide & des plus riches. f Des Élévations ſablonneuſes, lége

& Situation que ce ſoit.

a Evonimus Semper Virens.
c . . . . Latifolia.
b Vaccinium Arboreum.
. . . . . Pendulum.
d . . . . . Evonimifolium.
b . . . . . Puſillum.
. . . . . Racemoſum.
. . . . . Nigrum.
. . . . . Parvifolium.
d . . . . . Paluſtre.
c Rhododendron maximum.
Kalmia Latifolia.
d . . . Glauca.
. . . Anguſtifolia.
. . . Ciliata.
b Andromeda Arborea.
d . . . . . Plumata.
a . . . . . Spicata.
b . . . . . Lasifolia.
a . . . . . Lanceolata.
d . . . . . Nitida.
. . . . . Globulifera.
a . . . . . Graſſifolia.
c Lonicera Canadenſis.
c . . . . Periclimenum.
c . . . . Chamæceraſus.
b . . . . Symphoricarpos.
Diarvilla Canadenſis.
d Lythrum,
a Xanthoxilum Virginianum.
Glycine Fruteſcens.
. . . . Apios.
Oxyacantha Canadenſis.
c Betula Papyrifera.
a . . Lenta.
c . . Nigra.
a . . Nana.
c . . Populifolia.
d Alnus Rubra, Betula.
a . . Maritima.
. . Glauca.
c Acer Sachariflua.
a . . Rubra.
c . . Glauca.
. . Negundo.
c . . Arbuſtiva.
. . Ornata.
c Taxus Canadenſis.
Mitchelia Repens. Ombre.
Ariſtolochia Fruteſcens.
c Tilia.
a Fraxinus Excelſior.
c . . . . Nigra.
o Quercus Alba.
b . . . . Nigra.
. . . . Rubra.
. . . . Hiſpanica.
. . . . Nana.
a . . . . Phyllos.
. . . . Deltoide.
f . . . . Folio ampliſſima.
a . . . . Ægilops.
g . . . . Caſtanea.
. . . . Dentata.
b . . . . Chinquapin.
g . . . . Lobata.
b . . . . Campana.
o . . . . Paluſtris.
f . . . . Flamula.
. . . . Gallifera.
a Haleſia Tetraptera.
Ptelea.

far left, *Franklin tree, Franklinia alatamaha. William Bartram drawing engraved by James Trenchard, Philadelphia, ca. 1786.* left: *Facsimiles of 18th century seed packets.* above, *Seeds of Bartram plants over a facsimile letter from John Bartram to Peter Collinson.* right, *"Catalogue D'Arbres D'Arbustes et de Plantes…" Paris: 1783. The Bartram broadside catalogue of 1783 reprinted in Paris through the agency of Benjamin Franklin.*

1783

CATALOGUE of American TREES, SHRUBS and HERBACIOUS PLANTS, most of which are now growing, and
Seed in JOHN BARTRAM's Garden, near Philadelphia. The Seed and growing Plants of which are disposed of on the most reason

> It consists of a beautiful lawn, interspersed with little knobs or tufts of flowering shrubs, and clumps of trees, well disposed. Through the middle of the gardens, runs a spacious gravel-walk, lined with double rows of thriving elms, and communicating with serpentine walks which encompass the whole area.[34]

Vaughan compiled a list of seventy plants from fifty-five species he had purchased from the Bartram in April 1785 for the new garden. [35]

The Bartrams also hosted a procession of American and European visitors in the years following the Revolution. Johann David Schöpf, chief surgeon with the Ansbach troops in North America, visited Bartram's Garden several times during a brief stay in Philadelphia in the middle of 1783, writing:

> During the first days of my stay at Philadelphia, I visited among others Mr. Bartram, the son of the worthy and meritorious botanist (so often mentioned by Kalm) who died six years ago at a great age. Bartram the elder was merely a gardener, but by his own talents and industry, almost without instruction became first botanist in America, honored with their correspondence by Linnaeus, Collinson, and other savans.
> ...The son, the present owner of the garden, follows the employments of his father, and maintains a very respectable collection of sundry North American plants, particularly trees and shrubs, the seeds and shoots of which he sends to England and France at a good profit. He is not so well known to the botanical world as was his father, but is equally deserving of recognition. When young he spent several years among the Florida Indians, and made a collection of plants in that region; his unprinted manuscript on the nations and products of that country should be instructive and interesting. In the small space of his garden there are to be found assembled really a great variety of American plants, among others most of their vines and conifers, species of which very little is generally known.[36]

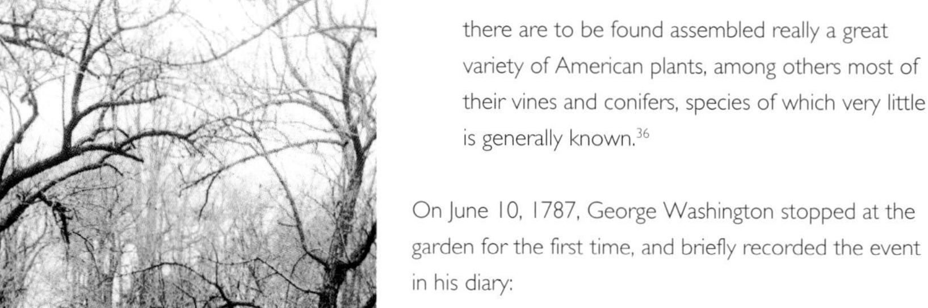

On June 10, 1787, George Washington stopped at the garden for the first time, and briefly recorded the event in his diary:

> Breakfasted by agreement at Mr. Powell's, and in Company with him rid to see the Botanical garden of Mr. Bartram; which, tho' Stored with many curious plts. Shrubs & trees, many of which are exotics was not laid off with much taste, nor was it large.[37]

Nevertheless, Washington returned on a number of occasions, and sought advice on live fencing from the Bartrams. He particularly admired the red cedar hedges on the Bartram farm and attempted to duplicate them in Virginia. Two large shipments of plants went from Bartram's Garden to Mount Vernon in 1792.[38]

The early morning visit of the Reverend Manasseh Cutler with members of the Constitutional Convention in July 1787 is one of the best-known descriptions of Bartram's Garden.

> This is a very ancient garden, and the collection is large indeed, but is made principally from the Middle and Southern States. It is finely situated, as it partakes of every kind of soil, has a fine stream of water, and an artificial pond, where he has a good collection of aquatic plants. There is no situation in which plants or trees are found but that they may be propagated here in one that is similar.[39]

Naturalists from home and abroad were frequent visitors to the Bartram collection, including André Michaux and his son Françoise André, and the American botanist

left, *William Bartram, "Catalogue of American Trees, Shrubs and Herbacious Plants, most of which are now growing, and produce ripe Seed in John Bartram's Garden, near Philadelphia".*
above, *Footbridge across the Philadelphia, Wilmington, and Baltimore Rail Road, looking west from the Bartram garden toward the orchard. From a set of nine views of Bartram's Garden by an unknown photographer, early 1892.*

Henry Muhlenberg first traveled to the Bartrams' garden from Lancaster, Pennsylvania, in 1792. Thomas Jefferson's first recorded trip to Bartram's Garden occurred in January 1783, although it is quite likely that he visited earlier.[40] He requested seeds from the Bartrams on multiple occasions while posted in Paris as the American minister, and again on his return to the United States. He maintained a sporadic correspondence with William Bartram, which probably under represents their friendship. Jefferson spent the summer of 1793 in a nearby house on the east bank of the Schuylkill, "in sight both Bartram's and Gray's gardens."[41]

William Bartram had a particularly long and close relationship with Benjamin Smith Barton, the first professor of natural history at the University of Pennsylvania. Barton had a practice of querying Bartram in detail on any number of topics of natural history, and much of the material Barton received from William Bartram ultimately found its way into his publications. Barton brought his classes to Bartram's Garden to study, and William Bartram expressed pride in his educational role, writing in the preface to the new catalogue issued in 1807 that the family garden: "may with propriety and truth be called the *Botanical Academy of Pennsylvania*, since, being near Philadelphia, the Professors of Botany, Chemistry, and Materia Medica, attended by their youthful train of pupils, annually assemble here during the Floral season."[42]

John Bartram Jr.'s wife, Elizah Howell Bartram (1751-1784), died young. John and his brother William were left to raise four small children. A series of misfortunes plagued the family in the beginning of the 19th century. John Jr. retired from most of his active business in 1800, and his son John Bartram III (1781–1804) assisted in running the botanic garden until his untimely death in 1804. A second son, James Howell Bartram (1783–1818), studying medicine with B. S. Barton was recalled home, but soon left on a voyage as a ship's surgeon to the Cape of Good Hope and Batavia.[43] The enlarged 1807 *Catalogue* of the garden was probably largely written by William Bartram. The list is headed "John Bartram & Son" and the introduction concludes: "Finding old age coming on, he has lately associated his son with him in the concern and hopes by their united exertions, the gardens will continue to be worthy of the attention of the lovers of science and the admirers of nature."[44] While the *Catalogue* was intended to launch the new business arrangements, the plans were ill-fated. A federal embargo on all foreign trade in response to the Napoleonic Wars effectively crippled the Bartrams' business. Although, the embargo was lifted in 1809, disruptions related to the European wars continued, and eventually drew the United States into the War of 1812. The American economy was in a state of collapse until at least 1815.

The Third Generation: Ann and Robert Carr

In the spring of 1809, the fortunes of the Bartram family garden improved with the marriage of Ann M. Bartram (Nancy) to Robert Carr, a Philadelphia printer. Ann had served as the female head of the Bartram household from 1794, at the young age of 15.

From 1802 through 1806, Ann and her uncle William tutored a Scots immigrant, Alexander Wilson (1766-1813), in drawing and ornithology. Wilson, a one-time weaver and a skilled poet, was schoolmaster at a one-room school situated at the northwest corner of John Bartram Jr.'s farm. Wilson spent much time with William Bartram and his niece Nancy at the garden, and was transformed by the experience. Wilson began collecting and drawing birds in the summer of 1803, and by 1806 had determined to produce a major illustrated work on birds. Wilson's *American Ornithology* brought Ann Bartram and Robert Carr together. Ann Bartram helped with the coloring, while Robert Carr and his brother William printed the work for the Philadelphia publishers Bradford & Inskeep.[45]

John Bartram Jr. died in November 1812, and Robert Carr and his brother-in-laws were named as executors of the estate, but the ongoing war prevented immediate settlement. Robert Carr served in the US Infantry during the war, and was not

Alexander Wilson, American Ornithology, Philadelphia: 1812, vol. 5, plate 42. Red Owl, Warbling Flycatcher, Purple Finch, Brown Lark.

above, *East facade of Bartram House, ca. 1882. Photograph by William H. Rau.*

Old cider mill on the bank of the Schuylkill. From a set of nine views of Bartram's Garden by an unknown photographer, early 1892.

Contemporary view of cider mill.

discharged until May 1815.[46] By late 1814 the family decided to divide the original Bartram farm into three tracts, and Ann and Robert Carr received the original Bartram house, botanic garden, and nursery, and the north meadow land, a total of thirty-two acres. (Roughly the bounds preserved in the current Bartram's Garden park). Robert and Ann Carr also came into ownership of "384 pots, boxes and tubs of plants" valued at 250 dollars, undoubtedly the plant stock at the garden at the time of John Bartram Jr.'s inventory.

Robert Carr's printing business was forced into sudden bankruptcy late in 1813 with the failure of several large publishing houses, in particular Bradford & Inskeep and W. P. Farrand and Co. This downturn came during production of Alexander Wilson's *American Ornithology*. Perhaps not coincidentally, Wilson died in Philadelphia on August 23, 1813, before completing the work's eighth volume.[47] With his printing business ruined, Carr returned from the war to begin a new career with his wife, Ann, as a horticulturist and nurseryman. Ann, alone of the Bartram children, wished to maintain the garden. Robert Carr might have preferred to sell or lease the garden property, but as he wrote in February 1819: "the advanced age of our uncle, Mr. W. B., who resides with us and who could not bear the thought of parting with the garden, forbids the idea of selling during his life."[48]

Ann B. Carr inherited the Bartram botanic talent and she was tutored from a young age by her Uncle William in drawing and painting from nature.[49] A description of the garden in 1837 by Alexander Gordon also praised her accomplishments:

> to speak in just terms respecting her enthusiasm for plants, (which is only equaled by her success in their cultivation) is a task I am incompetent to perform...the passionate fondness with which she toils among the plants, in every department, from the earliest dawn until darkness renders her operations impracticable.... Her knowledge of American plants is most extensive, not surpassed, if equaled, by anyone in the United States.[50]

The Carrs continued to cultivate within the framework of the botanic garden and plantings left by the first John Bartram, but it is clear they also enlarged the garden and broadened and increased the collection of plants. This generation of the Bartram family also issued printed catalogues of the botanic collection and in the 1820s and 1830s documented more than 2000 species of native and exotic plants available for sale.

Ann and Robert Carr were childless, but continued the family business with John Bartram Carr (1804–1839), a son from Robert Carr's first marriage, and a Carr nephew, George Gale (born ca. 1816).[51] Young Carr was probably also trained by William Bartram, and grew to adulthood surrounded by some of the most famous American naturalists of the day, including Thomas Say, Thomas Nuttall, William Baldwin, William Darlington, W. P. C. Barton, John Torrey, Richard Harlan, Constantine Rafinesque, and John James Audubon. In 1829, Carr attempted to secure his son a position as assistant botanist on a proposed polar expedition by the Navy Department. He described his son:

> as a **practical** botanist it will be difficult to meet with his equal, and he has been reared from infancy, in this garden, under the instruction of Mr. William Bartram, the botanist, a naturalist. With the exception of Mr. Nuttall, perhaps there is none better acquainted with the plants of this country, and his knowledge of exotics is very general.[52]

A good description of the Carr garden at its peak is preserved in the report of a committee appointed in the summer of 1830 by the Pennsylvania Horticultural Society to visit gardens and nurseries in the vicinity of Philadelphia. The committee concluded "in reference to out door plants" that the Bartram Botanic Garden: "must necessarily

above, *Lady Petre Pear tree at the southeast corner of the Bartram house. Started from seed by John Bartram in the 1730s. Photograph by Henry Dreer Bartram, 1898.*

stand unrivaled.... From this nursery, many thousands of plants and seeds are exported every season to Europe and South America. It is computed that there are 2000 species of our native productions, contained in a space of six acres."[53] Later in the decade, Alexander Gordon's visit of 1837 reveals that the nursery had grown to twelve acres, and there were ten specialized glass houses in the exotic department.[54]

Despite a rebound in business and an expansion in facilities in the 1820s and 1830s, the Bartram business ultimately failed. As North American plants became more widely cultivated, they were no longer rare or interesting to the gardening world. The Bartram family had depended on an elite group of customers in Europe and America, who were now interested in different plants.

Facing bankruptcy in 1850, Robert Carr was forced to sell the garden at auction.

> I regret to have to add that this place must now be sold.—My **poverty** but not **my will** consents. The business which supported this place so long (the cultivation of our native seeds and plants for Europe) has ceased to be worth attending to, and has been involving me in losses and debts for several years past.[55]

Some years later, John Jay Smith, librarian for the Library Company and editor of *The Horticulturist* after Andrew Jackson Downing, reminisced on the fate of the Bartram Garden:

> It was then the only example of various planting near us, and rich it was in trees and plants from far-off regions.... The habits of the Bartrams, when I knew them, were a continuance of the simplicity of preceding years. They still executed orders from Europe for seeds and plants, in a small way, but derived profit enough from the place for their moderate wants. I well remember the picking of the seeds of a fine redbud maple that stood, and probably still stands, near the old house, now not superseded in interest by the more pretentious mansion of the purchaser, Colonel Eastwick.... From this garden dates horticulture in America. It should be carefully preserved forever.[56]

Only a single eighteenth-century illustration is known of the Bartram house and garden, entitled A Draught of John Bartram's House and Garden as it appears from the River. This schematic view of garden in late 1758 is now generally thought to be by young William Bartram.

Following spread: 1- *Descendants gathered for the first Bartram family reunion, June 8, 1893. Photograph by Frederick Gutekunst.* 2 -*First floor Hall room of Bartram house following restoration work in 1926. Photograph by Philip B. Wallace.* 3- *West side of Bartram house and garden, ca. 1896-1900. Photograph by William H. Ingram.* 4- *Outbuildings at Bartram's Garden. Photograph by William H. Rau, 1895.* 5- *Kitchen of the Bartram house furnished as a sentimental family hearth, ca. 1900. Photograph by Philip B. Wallace.*

2
3
4
5

[1] Deed, Owen Owens, Sheriff to John Bartram, September 13, 1728, Philadelphia County Records, Deed Book GWC-41-356, Philadelphia County Records, Deed Book GWC–41–356.

[2] "The Bartram Women: Farm Wives, Artists, Botanists, and Entrepreneurs," *Bartram Broadside* (Winter 2001), p. 1; Edmund Berkeley and Dorothy Smith Berkeley, *The Life and Travels of John Bartram: From Lake Ontario to the River St. John* (Tallahassee: University Presses of Florida, 1982), p. 10.

[3] An edition Bartram's letters was published in 1992: *The Correspondence of John Bartram 1734–1777*, ed. Edmund Berkeley and Dorothy Smith Berkeley (Gainesville: University Press of Florida, 1992). For a biography of Collinson, see Norman G. Brett-James, *The Life of Peter Collinson* (London: Edgar G. Dustan & Co., 1925). A smaller selection of Collinson's correspondence is published in *'Forget not Mee & My Garden...' Selected Letters 1725–1758 of Peter Collinson, F.R.S.; Memoirs of the American Philosophical Society*, vol. 241, ed. Alan W. Armstrong (Philadelphia: American Philosophical Society), 2002.

[4] William Bartram, *Travels Through North & South Carolina, Georgia, East & West Florida...* (Philadelphia: James & Johnson, 1791). See the bibliography for the full title. Editions of *Travels* appeared in London in 1792 and 1794; Dublin in 1793; Berlin and Vienna in 1793; Haarlem between 1794 and 1797; Amsterdam in 1797; and Paris in 1799 and 1801.

[5] John Bartram to Peter Collinson, May 1, 1764, *Correspondence of John Bartram* 1992, p. 627.

[6] John Bartram to Sir Hans Sloane, Sept. 23, 1743, *Correspondence of John Bartram* 1992, p. 224.

[7] James Logan to Peter Collinson, June 8, 1736, The Historical Society of Pennsylvania (HSP), Logan Papers-Alverthrop Letterbook A, p. 4A.

[8] Logan to Collinson, August 20, 1737, HSP, Logan Papers-Alverthrop Letterbook A, p. 28B.

[9] Benjamin Franklin to his parents, Josiah and Abiah Franklin, Boston, September 6, 1744.

[10] "A Copy of the Subscription Paper, for the Encouragement of Mr. John Bartram," *Pennsylvania Gazette*, March 17, 1742, *Correspondence of John Bartram* 1992, pp. 188–189.

[11] Francis D. West, "John Bartram and the American Philosophical Society," *Pennsylvania History*, vol. 23 (Oct. 1956), pp. 463–466; Whitfield J. Bell Jr., *Patriot-Improvers: Biographical Sketches of Members of the American Philosophical Society*, vol. 1 (1743–1768), *Memoirs of the American Philosophical Society*, vol. 226 (Philadelphia: American Philosophical Society, 1997), pp. 48–49.

[12] The origin of this quote, a letter from Linnaeus to John Bartram of unknown date, is presumed lost. The quote was first recorded in an anonymous biography of Bartram from 1800 in the Philadelphia edition of the *Supplement to the Encylcopædia* (with William Bartram as the probable author). "Bartram, John," *Supplement to the Encylcopædia, or Dictionary of Art, Sciences, and Miscellaneous Literature* (Philadelphia: Printed by Budd and Bartram for Thomas Dobson, 1800), vol. 1, 91–92.

[13] Kalm's entry is dated September 25, 1748. Peter Kalm, *Travels into North America; Containing Its Natural History, and a Circumstantial Account of Its Plantations and Agriculture in General*, trans. John Reinhold Forster, 2nd edition, 2 vols. (London: 1772), vol. 1, p. 88.

[14] Bartram specimens can be found in the Sloane Herbarium at the Natural History Museum, London, and some have been identified in collections of the Oxford Botanic Garden. A large collection of Bartram herbarium specimens are preserved in three volumes (vols. XI, XII, and XIII) of Lord Petre's *Hortus Siccus*, at the Sutro Library, San Francisco, California.

[15] Joel T. Fry, "An International Catalogue of North American Trees and Shrubs: The Bartram Broadside, 1783," *The Journal of Garden History*, vol. 16 (Jan.–Mar. 1996), pp. 3–9. For details on Bartram's early clients and the impact of North American plants on English gardens, see Mark Laird, *The Flowering of the Landscape Garden: English Pleasure Grounds 1720–1800* (Philadelphia: University of Pennsylvania Press, 1999), especially chapter 2; and Douglas D. C. Chambers, *Planters of the English Landscape Garden: Botany, Trees, and the Georgics* (New Haven, Conn.: Yale University Press, 1993), chapters 6 and 7.

[16] Mark Catesby, *Natural History of Carolina, Florida and the Bahama Islands*, 2 vols. and appendix (London: 1731–48).

[17] John Bartram, *Observations on the Inhabitants, Climate, Soil, Rivers, Productions, Animals, and other Matters Worthy of Notice. Made by Mr. John Bartram, in his travels from Pensilvania to Onondago, Oswego and the Lake Ontario, in Canada* (London: 1751); *An Account of East-Florida, with a journal, kept by John Bartram of Philadelphia, Botanist to His Majesty for the Floridas; upon a journey from St. Augustine up the River St. John's*, ed. William Stork (2nd ed., London: 1767; 3rd ed., London: 1769; 4th ed., London: 1774).

[18] Berkeley and Berkeley, *Life and Travels of John Bartram*, pp. 226–228; Fry, "An International Catalogue," p. 6.

[19] *"A Draught of John Bartram's House and Garden as it appears from the River,"* Peter Collinson added the date *"1758"* and his own name, *"Sent to P Collinson."* The plan is housed in a volume of drawings once owned by Collinson and now located in the library of the Earl of Derby at Knowsley Hall. Dorothy T. Povey, "Garden of a King's Botanist," *Country Life* , vol. 119 (Mar. 22, 1956), pp. 549–550.

[20] [Henry Joel Cadbury], "The Disownment of John Bartram." *Bulletin of the Friends' Historical Association*, vol. 17, no. 1 (Spring 1928), pp. 16–22.

[21] John Bartram to Peter Collinson, May 30, 1763, *Correspondence of John Bartram* 1992, p. 594.

[22] John Bartram to Benjamin Franklin, April 29, 1771, *Correspondence of John Bartram* 1992, p. 739.

[23] Daniel L. Schafer, "The Forlorn State of Poor Billy Bartram," *El Scribano, The St. Augustine Journal of History*, vol. 32 (1995), p. 1-11.

[24] William Bartram, "Travels in Georgia and Florida, 1773–74: A Report to Dr. John Fothergill," ed. Francis Harper, *Transactions of the American Philosophical Society*, vol. 33 (Nov. 1943), pp. 121–242.

[25] Francis Harper, "Proposals for Publishing Bartram's *Travels*," *American Philosophical Society Library Bulletin* (1945), pp. 27–38.

[26] Bartram, *Travels.* A number of works have traced the influence of Bartram's *Travels* on Coleridge, Wordsworth, Chateaubriand, and other European writers of the Romantic period. See John Livingston

Lowes, *The Road to Xanadu: A Study in the Ways of Imagination*. (New York: Houghton Mifflin, 1927), and N. Bryllion Fagin, *William Bartram: Interpreter of the American Landscape*. (Baltimore: Johns Hopkins Press, 1933).

[27] William Bartram to Benjamin Smith Barton, undated draft, ca. Aug. 26, 1787 or Feb. 19, 1788, Item 104A, Jane Gray Autograph Collection, Archives of the Gray Herbarium, Harvard University, Cambridge, Mass.; Dr. Autenrieth to William Bartram, November 9, 1795, Bartram Papers 1:1, HSP; William Bartram to Lachlan McIntosh, May 31, 1796, New York Historical Society, Misc. Mss. Bartram.

[28] Fry, "An International Catalogue," 9–14.

[29] Joseph Matthias Gérard de Rayneval, or Sieur Gérard, the first French minister to the new government of the United States, arrived in Philadelphia July 12, 1778, and remained on duty until ill health required his return to France in October 1779. *"Catalogue of Roots & Seeds in a Small Box for his Exelly the Minister of France" and "Catalogue of Roots in two Boxes for his Excelly the French Minister,"* Bartram Papers 4: 106, HSP.

[30] John and William Bartram to Carl Linnaeus Jr., August 16, 1783, Uppsala University Library, Ms UUB G 359 (Bartram), Uppsala, Sweden.

[31] Benjamin Franklin to John Bartram, May 27, 1777, *Correspondence of John Bartram* 1992, p. 771.

[32] [William Bartram], *CATALOGUE of American TREES, SHRUBS and HERBACIOUS PLANTS, most of which are now growing, and produce ripe Seed in John Bartram's Garden, near Philadelphia....* ([Phila.]: [1783]); *CATALOGUE D'ARBRES D'ARBUSTES ET DE PLANTES....* (Paris, 1783).

[33] Sarah P. Stetson, "The Philadelphia Sojourn of Samuel Vaughan," *Pennsylvania Magazine of History and Biography*, vol. 73 (Oct. 1949), pp. 459–474.

[34] *The Columbian Magazine*, (Jan. 1790), pp. 25–26.

[35] *"Planted in the State-house square,"* and Samuel Vaughan to Humphry Marshall, May 28, 1785, series X, manuscripts, box 10/4, file "Humphry Marshall Papers," United States Department of Agriculture History Collection, Special Collections, National Agricultural Library, Washington, DC.

[36] Johann David Schöpf, *Reise durch einige der mittlern und südlichen Vereinigten nordamerikanischen Staaten nach Ost-Florida und den Bahama-Inseln unternommen in den Jahren 1783 und 1784*, 2 vols. (Erlangen: 1788); *Travels in the Confederation, 1783–1784*, trans. Alfred J. Morrison (Philadelphia: William J. Campbell, 1911), vol. 1, 90–93.

[37] *The Diaries of George Washington*, ed. Donald Jackson and Dorothy Twohig (Charlottesville: University Press of Virginia, 1979), vol. 5 (July 1786–December 1789), pp. 166–167.

[38] *"Catalogue of Trees, Shrubs, & Plants, of Jnº. Bartram,"* Mar. 1792, George Washington Papers, Library of Congress (LOC), printed in Philander D. Chase, *The Papers of George Washington: Presidential Series* (Charlottesville: University of Virginia Press, 2002), vol. 10, pp. 175–183; list of ninety-seven species sent from Bartram's Garden to Mount Vernon for George Washington, in hand of Bartholomew Dandridge, Washington's secretary, November 7, 1792, George Washington Papers, LOC.

[39] Manasseh Cutler, July 1787, in *Life, Journals, and Correspondence of Rev. Manasseh Cutler, LL.D*, 2 vols, ed. William Parker and Julia Perkins Cutler (Cincinnati: 1888), vol. 1, pp. 272–273.

[40] *Jefferson's Memorandum Books, Accounts, with Legal Records and Miscellany, 1767–1826,* The Papers of Thomas Jefferson Second Series, vol. 1, ed. James A. Bear Jr. and Lucia C. Stanton (Princeton, NJ: Princeton University Press, 1997), p. 526.

[41] Thomas Jefferson to Martha Jefferson Randolph, May 26, 1793, *The Papers of Thomas Jefferson*, vol. 26, (Princeton, NJ: Princeton University Press, 1995), 122.

[42] [William Bartram], A Catalogue of Trees, Shrubs, and Herbaceous Plants, Indigenous to the United States of America; Cultivated and Disposed of By John Bartram & Son, At their Botanical Garden, Kingsess, near Philadelphia (Philadelphia: 1807), p. 7.

[43] James Howell Bartram, *"Remarks and Observations on A Passage to the C. of Good H., Batavia, Madras, Calcutta, &c., &c., by James Bartram, Surgeon, A. D. 1804 & 1805,"* printed in facsimile with biographical sketch by Joel T. Fry (Philadelphia: The John Bartram Association, 1991).

[44] [William Bartram], *A Catalogue of Trees, Shrubs, and Herbaceous Plants*, 1807, p. 7.

[45] Clark Hunter, ed., *The Life and Letters of Alexander Wilson* (Philadelphia: American Philosophical Society, 1983), pp. 74, 82, 89. Wilson spent summers of 1809–12 boarding with the Carrs at Bartram's Garden while working on *American Ornithology*.

[46] William Bartram Snyder, *"Biographical Sketch of Colonel Robert Carr,"* Read before the HSP, December 10, 1866, p. 9; Robert Carr, "Journal and Letter Book, Etc.," 1811–23, MS volume, HSP.

[47] Vols. 8 and 9 of *American Ornithology* were issued posthumously in 1814, and edited by George Ord.

[48] Carr, "Journal and Letter Book."

[49] Snyder, "Biographical Sketch of Colonel Robert Carr," pp. 22–23.

[50] Alexander Gordon,"Bartram Botanic Garden," *The Genesee Farmer* vol. 7, no. 28 (July 15, 1837), p. 220.

[51] Joel T. Fry, "John Bartram Carr: The Unknown Bartram," *Bartram Broadside* (Fall 1994), pp. 1–9.

[52] Robert Carr to General T. D. Barnard, Jan. 23, 1829; and R. Carr to Secretary of the Navy, Jan. 23, 1829, both Chester County Historical Society, Chester County, Pennsylvania.

[53] "Report of the Committee appointed by the Horticultural Society of Penn'a, For visiting the Nurseries and Gardens in the vicinity of Philadelphia—13th July, 1830," *The Register of Pennsylvania*, ed. Samuel Hazard, vol. 7 (Feb. 12, 1831), pp. 105–106.

[54] Gordon, "Bartram Botanic Garden," 220.

[55] Robert Carr to William Darlington, April 10, 1850, Darlington Papers, New York Historical Society, New York, New York.

[56] John Jay Smith, *Recollections of John Jay Smith Written by Himself. Edited by his daughter, Elizabeth P. Smith* (Philadelphia: J. B. Lippincott, 1892), pp. 275–276.

TRAVELS

THROUGH

NORTH AND SOUTH CAROLINA,

GEORGIA,

EAST AND WEST FLORIDA,

THE CHEROKEE COUNTRY,

THE EXTENSIVE TERRITORIES OF THE MUSCOGULGES

OR CREEK CONFEDERACY,

AND THE COUNTRY OF THE CHACTAWS.

CONTAINING

AN ACCOUNT OF THE SOIL AND NATURAL PRODUCTIONS OF THOSE REGIONS;

TOGETHER WITH

OBSERVATIONS ON THE MANNERS OF THE INDIANS.

EMBELLISHED WITH COPPER-PLATES.

By WILLIAM BARTRAM.

PHILADELPHIA: PRINTED BY JAMES AND JOHNSON. 1791.

LONDON:

REPRINTED FOR J. JOHNSON, IN ST. PAUL'S CHURCH-YARD.

1792.

"At the request of Dr. Fothergill, of London, to search the Floridas, and the western parts of Carolina and Georgia, for the discovery of rare and useful productions of nature, chiefly of the vegetable kingdom; in April 1773, I embarked for Charleston, South-Carolina."
— William Bartram Travels, 1791

1773

March 20: sails from Philadelphia.

March 31: arrives in Charleston, South Carolina.

Mid-April: travels to Georgia coast, first sailing to Savannah then exploring overland to Midway, Sunbury, Darien, and Brunswick, and the Sea Islands.

April 24 or 25: crosses Altamaha River at Fort Barrington, re-encounters Franklinia which he and his father discovered in 1760.

May 1: after spending time with the McIntosh family, leaves Darien with young John McIntosh for Savannah and Augusta, intending to attend to Indian congress in Augusta.

May 5: at Ebenezer, Georgia.

May 9: visits George Galphin at Silver Bluff on the Savannah River.

May 14: at Augusta, Georgia. In ensuing days visits Wrightsborough.

June 3: congress of colonial officials with Creeks and Cherokees concludes with the second treaty of Augusta, and a large cession of native lands.

June 7: Bartram joins party to survey north boundary of New Purchase Cession. Travels north up the Little, Broad, Oconee, and Tugaloo Rivers to the Great Buffalo Lick and Cherokee Corner.

mid-July-end of year: returns south along Savannah River to Augusta and Savannah. Remains on Georgia coast through March 1774, exploring Sea Islands and Altamaha River. Considerable time spent at home of Lachlan McIntosh in Darien recovering from fever. Prepares collections, specimens, drawings, and plants for shipment from Savannah to Dr. Fothergill in London.

1774

March: begins travels to Frederica and Cumberland Island, Georgia, and Amelia Island, Florida, then via canoe to Cow-ford at current Jacksonville.

Mid-April: explores up St. Johns River past Fort Picolata and Rollestown to Spaldings Lower Store near Palatka.

Late-April: travels overland to Cuscowilla and the Alachua Savannah.

Mid-May to June: continued explorations on St. John River to Mount Royal, Lake George, and Spaldings Upper Store, encounter with alligators at Lake Dexter, and visit to Blue Springs, then returning to Spaldings Lower Store.

Mid-June to July: returned to Alachua with a company of traders, then across Florida to the Suwannee River, the Seminole town of Talahasochte, and Manatee Springs, then back to Spaldings Lower Store.

August-September: further exploration of St. Johns River to the upper store, experiences a hurricane near Beresford Plantation, much time at Spalding's Lower Store.

Early November: leaves Florida for coastal Georgia, November 10th at Broughton Island.

November-February: prepares specimens, journals, drawings, plants for shipment to Dr. Fothergill in London.

1775

Late March: Bartram's specimens and journals, and several boxes of plants sail on Henry Laurens' rice ship for England to the Isle of Wight, to be forwarded to Dr. Fothergill.

March 25, 1775: Bartram travels back to Charleston with Henry Laurens.

April 22: leaves for Cherokee country, initially up Savannah River.

Early May: travels from Augusta to Fort James.

May 10: leaves Fort James, travels to Loughabber, plantation of Alexander Cameron.

May 15: at Seneca and from there to Fort Prince George on the Keowee River.

May 19: departs Fort Prince George, crosses Oconee Mountain, to Chattooga River, follows Warwoman Creek, then north over divide and along Little Tennessee River, visiting Cherokee Middle towns.

May 22: arrives at Cherokee town Cowee, and explores Cowee Mountains.

May 24: leaves Cowee and crosses Nantahala Mountains at Burningtown Gap, toward the Cherokee Overhill towns. Encounters Attacullaculla on the road, soon after turns back east.

May 27: returns to Cowee.

May 30: returns to Keowee and Fort Prince George.

Early June: at Fort James, explores the Broad River.

June 22: at Fort Charlotte joins party of traders and adventurers bound for Mobile in West Florida. Cross Savannah north of Augusta and enter the Lower Creek Trading Path.

June 27: at Flat Rock, then crosses Ogeechee River.

July 1: at Rock Landing on the Oconee River.

July 3: crosses the Ocmulgee River.

July 5: discovers oakleaf hydrangea, crosses the Flint River.

July 11: arrive at Yuchi Town on the Chattahoochee River, and Creek town Apalachicola.

July 13: leaves the Lower Creek towns for 3 day journey to Tallassee on Tallapoosa River.

July 16: travels from Tallassee through Atasi to Kolumi.

July 19: in the Alabama prairies, observes rosinweed.

July 20 crosses Pintlalla Creek.

July 21: crosses Escambia River, travel through large savannahs and cane meadows.

Around July 29: arrives at Major Robert Farmar's plantation on the Tensaw River, modern Upper Bryant Landing near Stockton, AL.

July 30: down Tensaw delta to Mobile Bay, arrives at Mobile or Fort Condé.

August 5: returns up Tensaw delta to Major Farmar's plantation at Tensaw Bluff.

Mid-August: several explorations up Tensaw delta, discovers largeflower evening primrose, explores north to confluence of Tombigbee and Alabama Rivers, begins to feel symptoms of a dangerous fever.

End of August: returns to Mobile.

September 3-4: short sailing trip to Pensacola.

September 6: returns to Mobile, becomes severely ill with fever.

September 8: leaves Mobile by water for the Mississippi River.

September 9: ill, Bartram stops at home of French planter on Pearl River.

September 11-mid-October: close to death, Bartram seeks medical treatment at plantation of James Rumsey on Pearl Island. Remains there to recuperate for almost a month.

Early October: coasts along north shore of Lake Pontchartrain to Lake Maurepas.

October 21: reaches British colonial settlement at Manchac and Mississippi, travels north on Mississippi with William Dunbar to plantation near Baton Rouge, visits Alabama Town.

October 27: with Dunbar, visits White Cliffs and Pointe Coupée.

November 10: leaves Baton Rouge for return down Mississippi and then eastward.

November 16: returns to Mobile. Packs and ships collections to Dr. Fothergill via the merchants Swanson and M'Gillavry.

November 27: leaves Mobile for return to east coast with company of traders.

November 28: takes leave of Major Farmar at Tensaw.

December 4: arrives at Tallapoosa River, Creek town Mucclasse, visits Alabama, site of Fort Toulouse.

Passes through Creek towns, Kolumi, Tuckabahtchee, Atasi.

1776

January 2: leaves Upper Creek Towns, crosses Chattahoochee River, passes through Ocmulgee.

January 14: at Augusta.

Late January: at Savannah.

Spring & Summer: explores Georgia coast, travels to the St. Mary's River.

July: volunteers with Georgia Militia under Lachlan McIntosh during brief military campaign to dislodge loyalists from the St. Mary's River.

Summer: explores Altamaha River by canoe, this or later trip to collect seed of Franklinia and Pinckneya.

Late October: leaves Darien for Savannah, possible final shipment of specimens and plants to Fothergill.

Early November: leaves Savannah.

Mid-November: few days in Charleston.

Early December: at Ashwood plantation, Cape Fear River, NC.

December 12 or 13: leaves Ashwood for return home.

December 26: arrives at Alexandria, VA.

Late December: crosses Susquehanna on ice to Lancaster, PA.

1777

Around January 2: returns home to Bartram's Garden.

William Bartram - A Portfolio

"Motacilla. A Sollaterry Bird," Louisiana Waterthrush, Seiurus motacilla, *and "Yellow Spiked Lycimacha," swamp candles,* Lysimachia terrestris. *One of Bartram's earliest works, watercolor & ink on paper, early 1750s. The note at the bottom is by the British naturalist George Edwards.*

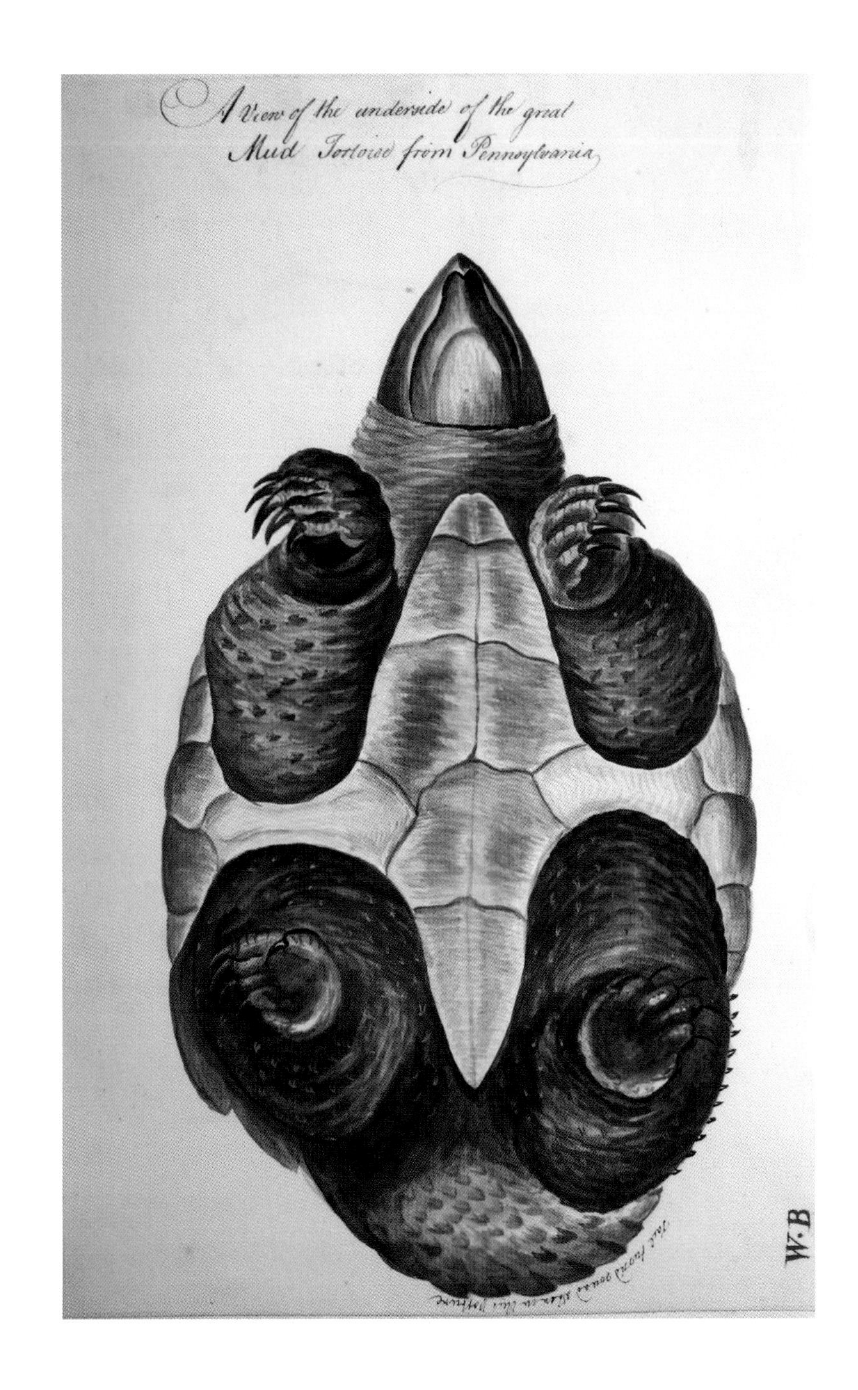

"A View of the underside of the great Mud Tortoise from Pennsylvania." Common snapping turtle Chelydra serpentina, *watercolor & ink on paper, 1759.*

"Purple Flower'd Ixia" from Florida, the celestial lily, Calydorea coelestina, *watercolor & ink on paper, 1767.*

"The great Silver-Leafed River Maple," silver maple, Acer saccharinum, *from a set of all the Pennsylvania maples sent to Peter Collinson, watercolor & ink on paper, 1755.*

Two Florida plants, gopher apple, Licania michauxii, *and hairy laurel,* Kalmia hirsuta, *drawn by William Bartram and engraved by James Trenchard, Philadelphia, 1786.*

Large whorled pogonia, Isotria verticillata, *and rosebud orchid,* Cleistes divaricata, *ink drawing and description of two North American orchids from 1796. The image is decorated with a Venus's flytrap and sundew in the foreground, and a landscape view of the city of Philadelphia at the right.*

"The Great Alachua-Savana; in East Florida," map and nature tableau of this unique environment, created during or after his visit in 1774.

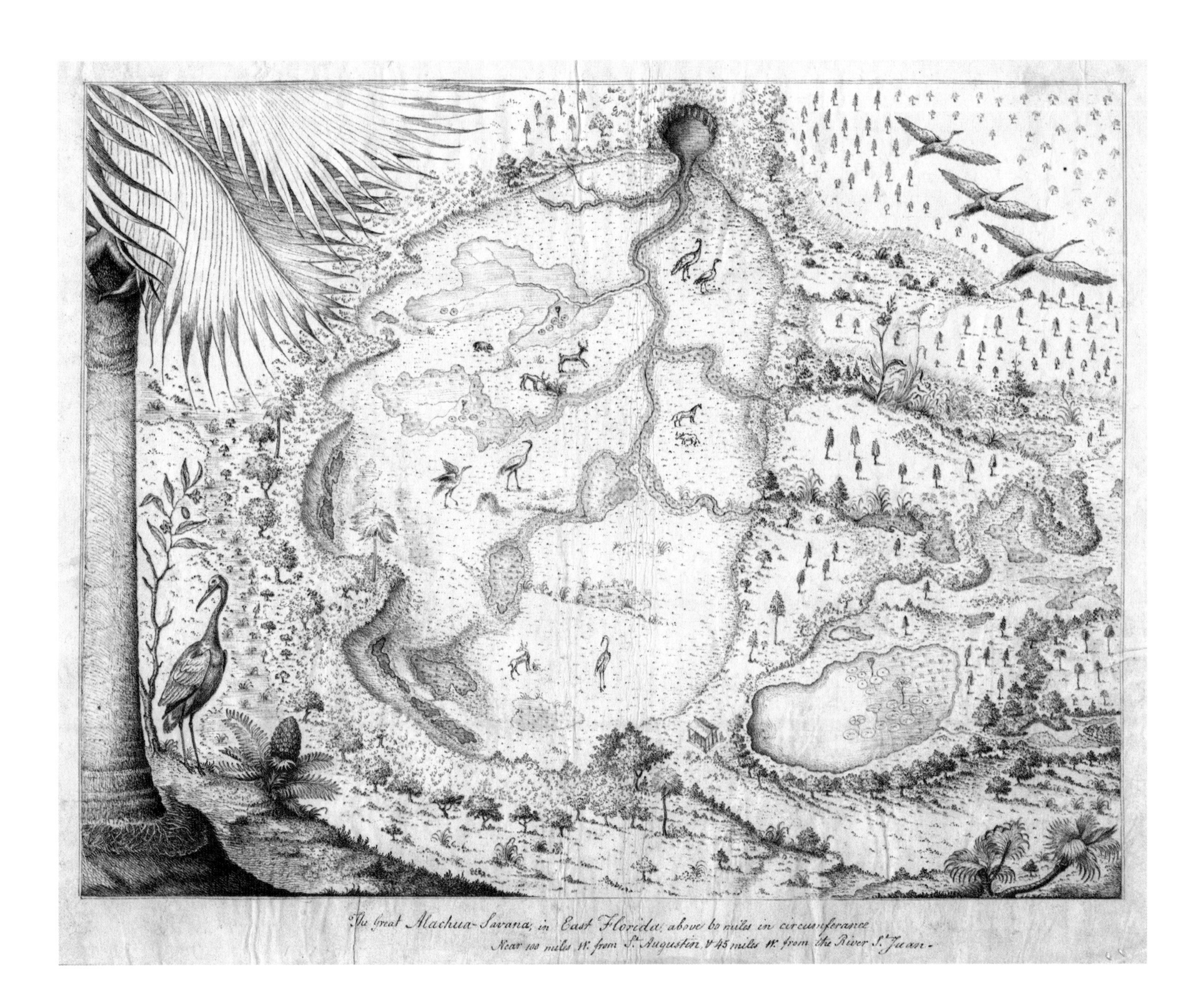
The Great Alachua-Savana, in East Florida, above 60 miles in circumferance
Near 100 miles W. from St. Augustin & 45 miles W. from the River St. Juan.

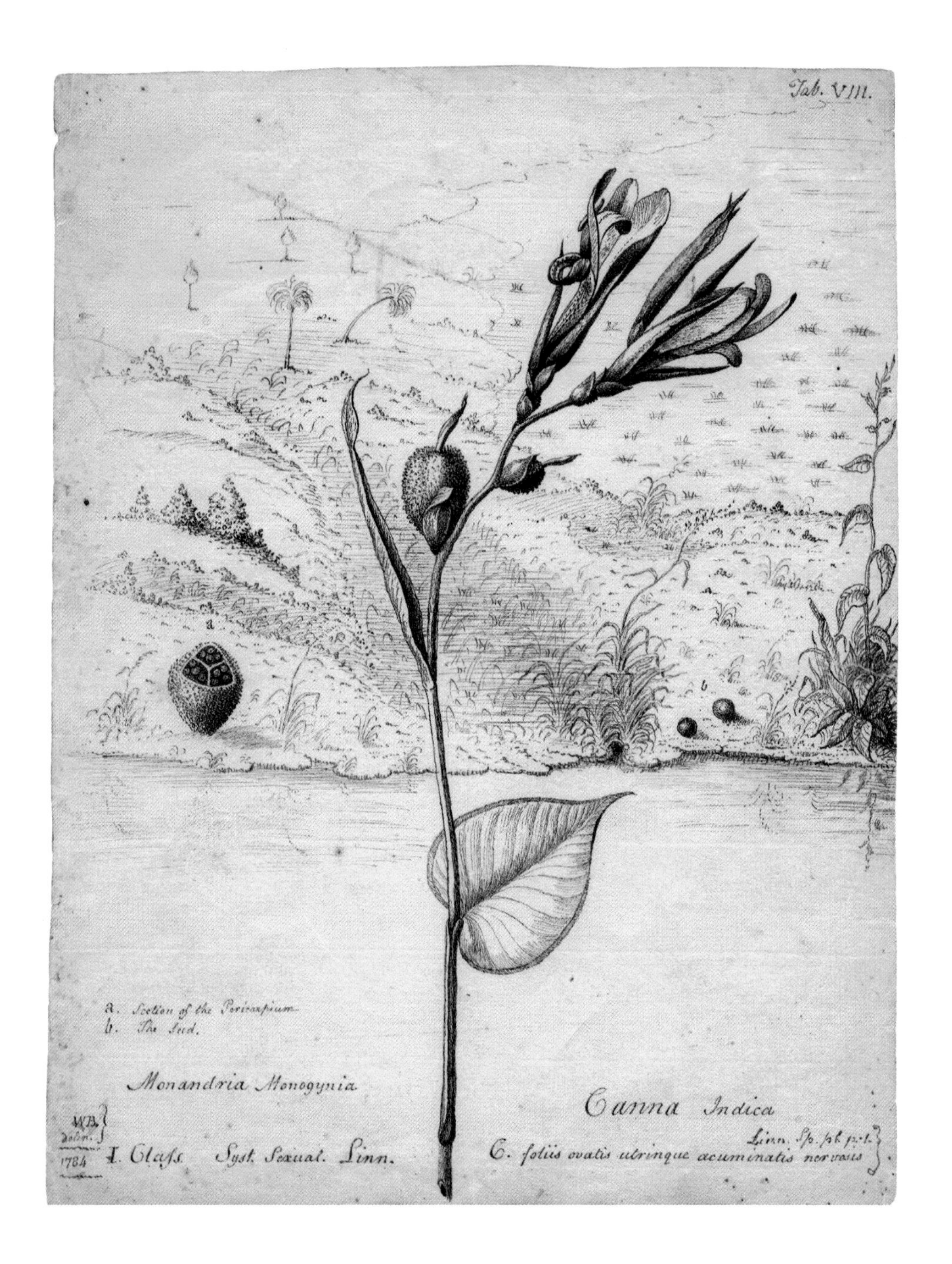

Indian shot, Canna indica, *drawing from 1784. The flowering specimen, with Linnaean class and description is floating before a Florida landscape.*

Fevertree, Pinckneya bracteata, *William Bartram drawing, engraved by James Trenchard, Philadelphia, 1786.*

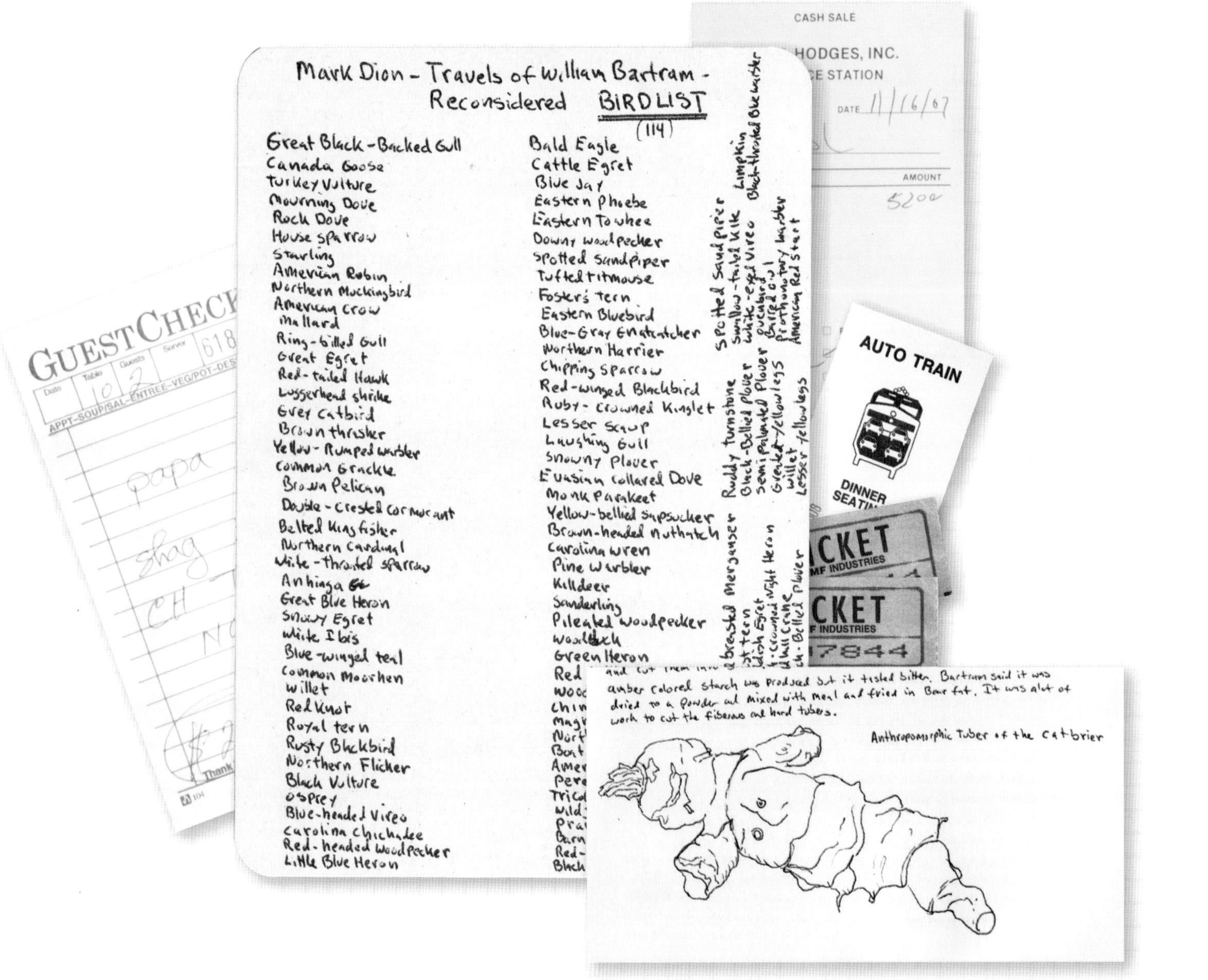

Mark Dion - Travels of William Bartram - Reconsidered BIRDLIST (114)

Great Black-Backed Gull
Canada Goose
Turkey Vulture
Mourning Dove
Rock Dove
House Sparrow
Starling
American Robin
Northern Mockingbird
American Crow
Mallard
Ring-billed Gull
Great Egret
Red-tailed Hawk
Loggerhead shrike
Grey Catbird
Brown thrasher
Yellow-Rumped warbler
Common Grackle
Brown Pelican
Double-crested cormorant
Belted Kingfisher
Northern Cardinal
White-throated sparrow
Anhinga
Great Blue Heron
Snowy Egret
White Ibis
Blue-winged teal
Common Moorhen
Willet
Red Knot
Royal tern
Rusty Blackbird
Northern Flicker
Black Vulture
Osprey
Blue-headed Vireo
Carolina Chickadee
Red-headed Woodpecker
Little Blue Heron

Bald Eagle
Cattle Egret
Blue Jay
Eastern Phoebe
Eastern Towhee
Downy woodpecker
Spotted Sandpiper
Tufted titmouse
Foster's tern
Eastern Bluebird
Blue-Gray Gnatcatcher
Northern Harrier
Chipping Sparrow
Red-winged Blackbird
Ruby-crowned Kinglet
Lesser Scaup
Laughing Gull
Snowy Plover
Eurasian collared Dove
Monk Parakeet
Yellow-bellied Sapsucker
Brown-headed nuthatch
Carolina wren
Pine Warbler
Killdeer
Sanderling
Pileated Woodpecker
Woodcock
Green Heron
Red
Wooc
Chin
Magr
Nort
Boat
Amer
Pere
Trico
Wild
Prai
Barn
Red-
Black

...breasted Merganser
...st tern
...dish Egret
...-crowned Night Heron
...hill Crane
...ck-Bellied plover

Ruddy turnstone
Black-Bellied plover
Semipalmated Plover
Greater Yellowlegs
Willet
Lesser Yellowlegs

Spotted Sandpiper
Swallow-tailed Kite
White-eyed Vireo
Ovenbird
Barred owl
Prothonotary warbler
American Red Start

Limpkin
Black-throated Blue warbler

CASH SALE
HODGES, INC.
...CE STATION
DATE 11/16/07
AMOUNT
5200

AUTO TRAIN
DINNER SEATI...

...CKET
...MF INDUSTRIES
...CKET
...F INDUSTRIES
...17844

GUESTCHECK
Date | Table 10 | Guests 2 | Server 618
APPT-SOUP/SAL-ENTREE-VEG/POT-DES...
papa
shag
CH
N...
Thank

and cut them into ... amber colored starch was produced but it tasted bitter. Bartram said it was dried to a powder and mixed with meal and fried in Bear fat. It was alot of work to cut the fiberous and hard tubers.

Anthropomorphic Tuber of the Catbrier

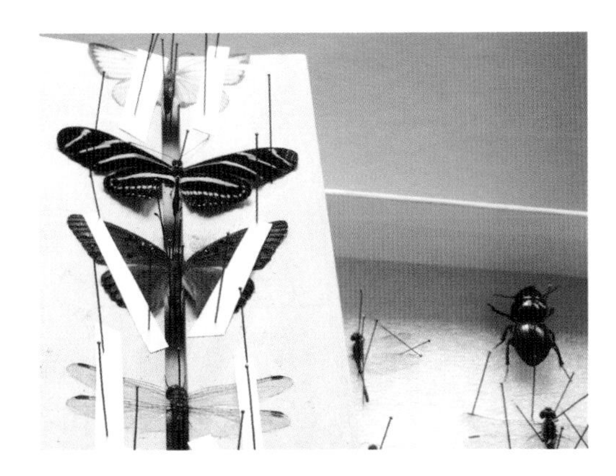

May 2nd, 2008
Days at the Atlantic Center for the Arts

After meeting with Nancy and Barbara I spent a little time with the butterfly net. I captured and got a number of dragonflies and butterflies.

Then the entire crew of art associates left for a lunch at JB's Fish Camp. On the drive over Vincent and Filip interrogated me about gallery business. After a cheerful lunch we loaded up for the Canaveral Seashore beach with its turbulent waves. We made a strange sight with Filip in his Euro Speedo and Bronwen helping recover treasures. We passed flocks of Ruddy Turnstones. In the surf numerous Portuguese man-of-war had washed up looking like oily balloons. I collected a few. Nancy seemed particularly interested in them. After about two and a half hours we returned to ACA for a last meal and open studios.

About fifty people showed up from the public. Ann gave an invigorating opening speech and spoke of how Florida has gone from being the 4th largest supporter of arts funding to being the 38th. Then David presented a work and Omar and another fellow presented. The Susan Marshall's associates danced and she showed her dancers in milk clip. I then asked my associates to line-up and introduce themselves with a few words. Although they were nervous they did a remarkable job. There was a rather silly collaboration between a composer and two dancers and then my group dressed like funky athletes and played a ping-pong championship. The night and residency ended with an absurd dance party in which Corrine and Filip were real stars. Two of the world's most uninhibited dancers acting out. Corrine in Dana's 1950 bathing suit and Filip in shorts and a wife beater. David turned to me and said "how long do you think we have to stay?" and we left just then.

May 4th, 2008
Homeward Bound

We had pleasant breakfast with Adele in her spotless condo. After a tour of the common areas and swimming pool we set off for the flea market, which was one of these labyrinthine covered affairs selling mostly cheap hardware and mass produced goods. We did find one or two items of interest like a push button doorbell from the 1920's. We then tried to visit a fireworks store which was closed but we did manage to come across a tiny antique shop were a few items were added to our obscurely packed car.

After bidding Adele goodbye we sped across the state to Sanford where the car train awaited us. Dana became obsessed with having a taco and we had to pull off the highway several times before we came across a Chipotle restaurant. Soon we were resting in our sleeping room on the Amtrak car train, which cost almost twice as much as it did last time. On the way to the station I spied a Northern Harrier hawk by the roadside, which made for my 80th and last bird on the trip.

It took about three and a half unhappy hours to box everything and cram it into the car. The vehicle is jammed and still I left behind my box of books, tools and writing box and butterfly nets.

---

next page, *1- Fontainebleau State Park; 2-,4-,10- Charleston, SC; 3-,9- New Smyrna Beach, FL; 5- Disney World Orlando, FL; 6- Spruce Creek Preserve, FL; 7- Mosquito Bay, FL; 8- Mississippi River, New Orleans, LA;*

I

2
3
4
5
6
7
8
9
10

Bartram Canoe Triangle, December 1-3, 2007

"Thus far we were finding it rather challenging to encounter wild places. Many of the Atlantic coastal areas seem subsumed by the canker of new development, while the coastal park swamps of Louisiana and Mississippi remain largely closed due to recent catastrophic events, and the Gulf Shores State Park is a mere postage stamp. Thus we had high hopes for Alabama's Bartram Canoe Trail and the river delta did not disappoint in providing high adventure.

"We met out traveling companion, the artist Christy Gast, in Pensacola. After I lectured at the University of West Florida and we feasted on local oysters we were ready for a hearty canoe trip. Generous friends of Christy's lent us a spacious and pristine red canoe, and by 10 a.m. we had already set off for a day paddle and overnight sleeping platform somewhere. Having consulted the glossy bold fold out brochure and slick web site we felt confident of a relaxing trip. The only indication that we would have some difficulty came from an elder waterman who claimed that no one had gotten through the creek we planned to paddle since the hurricane.

"With no further indication that things might get hairy we set off with a full canoe in high spirits and followed the canoe trail's prominent yellow diamond signs. The day bright and temperate, we had a good paddle ahead of us but it did not seem beyond our means. Soon however the reliable signs gave out and we began to see the havoc hurricanes had dealt the Delta. Our coordinates came into doubt as the trail signs petered out. At our first vital crossroad we had to choose between an impossible tangle of fallen trees and debris from what must have been a walkway or bridge on our left and a creek mouth blocked by a single massive tree fallen across the creek on our right. Submerged below the amber water we could faintly detect a trail marker with an arrow pointed down to unplumbed depths. We took the left fork, which meant portaging all our gear and canoe around the tree. This took so much time and muscle that we immediately devoured our fried chicken lunch as soon as the task was accomplished. We then perceived that the creek dead-ended into a road not 20 yards beyond. We reverse portaged and took the other fork, which entailed clearing the flotsam and jetsam and hauling the loaded canoe over a log that blocked our path. This creek also seemed utterly blocked and dead ended. At this point in dire need to consult the map we came to the realization that the state game lands map we possessed was of a scale too small to be useful, for it showed everything in microscopic detail without differentiation as to the size of the waterways, while the glossy Bartram

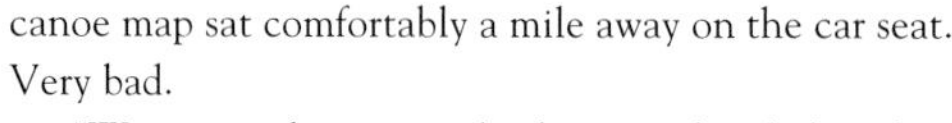

canoe map sat comfortably a mile away on the car seat. Very bad.

"We retraced out route back to another fork and followed what we now know to have been Major's Creek. This creek, with its high clay banks and sandy bottom, snaked back and forth for some time before we reached our first seriously obstructed section. Before us, stretching across the creek like massive pick-up-sticks were trees of various sizes fixed in every possible angle. For miles on end we had to haul the canoe over logs standing mid-stream, perched on mud and slime covered trunks, pulling the fat, red boat over the prone trees. At other times we would have to do a spine-bending limbo dance as we glided under low-hanging trees, or we would have to hack our way around root wads with only the leatherman saw blade to aid us. The car was full of tools to aid in such an endeavor - there were camp axes, bow saws and pruners, but we had anticipated an easy canoe trip. A cakewalk, not something hardcore where we would have to portage or haul the canoe every ten yards. Already soaked we waded through the creek dragging the canoe along over six inch deep water through sandbars and then again over three foot high fallen trees, all the while conscious of time - we began to imagine having to pitch camp.

"After hours at this arduous pace the river began to widen. Our map was soaked and in tatters. We had become so disoriented that it mattered little. Down this new, generous waterway we vigorously paddled certain that at any moment some vital clue would reveal our whereabouts. The large river abruptly ended at an elevated sandbar and we followed a narrow channel, which opened to our right. This shallow, twelve foot wide stream bisected the sandbar and palmetto bordered forest. After bottoming out several times the channel opened into an expansive river. Was this the Tensaw at last?

"The sun sets early and swiftly this time of year in southern Alabama and the blue of nightfall with its quiet was firmly establishing itself on the big river as we entered. Far in the distance Dana spied a small bass boat with two anglers. We had been paddling for eight hours interrupted only by the more physically challenging task of freeing the canoe and yet we put in steam and rushed toward the fishermen. They were probing the overhanging roots and cypress buttresses with skill and approached us as we zeroed in to them like hungry ducks on the slope. Our first question to them was "where are we?" Their first question to us was "where are you from?"

"No, we were not on the Tensaw River. We had gone far northwest and were now on the Alabama River. Bad news. They towed us back the half mile we had come to meet them to the sandbar. The anglers seemed shocked by the mess we had gotten

left, *Fragments of map which blew overboard*; above, *Paddling and hopelessly lost on the Bartram Canoe Trail.*

ourselves into - just minutes before nightfall, utterly lost and seemingly unprepared. However, we impressed them by mentioning our t-bone steak, green beans, tortellini, coffee, red wine and single malt scotch. They were charming gentlemen, hospitable and kind. They offered to tow us out and help us, but we were determined to stay. They left us with the highest possible esteem for the sportsmen of this fine state. As their running light disappeared around the bend we discovered that I had mistakenly not packed the tent. Me, experienced expedition camper, had left the tent and the map. Perhaps, I should cut back on the booze.

"We built a significant fire and prepared to sleep under the stars. We ate a simple meal, which could not have tasted better. After dinner we set a 17 hook night line baited with stinky orange chicken livers. This turned out to be a waste of time. Throughout the night our sleep was interrupted by the most outrageous wild sound. The mad screams of the great blue heron, owl hoots, critters rambling in the dry leaves and alligator calls. The morning came gray and gloomy and revealed strewn about us the tidings of every slob who had ever camped here. Scattered Busch beer cans, broken lawn chairs, entire full black plastic trash bags punctuated the landscape. Nevertheless the dawn chorus was glorious and fish rose from the river as an otter swam by. Christy caught a good size fish on a daredevil lure, and Dana caught a microscopic bass. I caught nothing. We ate fish, eggs and grits and were reluctant to leave our peaceful campsite. We loaded into the canoe and began to follow the instructions of our angler friends down river, past the cow stile, below the high tension wires and under the bridge. We hauled the boat over five exceptionally large trees and portaged the entire boat and contents once. During the portage I walked face first into an impressive yellow spider web with a three inch long beast in its center. She now dwells in a bottle of alcohol.

"It seemed we were on track and even our horrible little map assisted us until it blew overboard, becoming as flimsy as a wet napkin. Then something went wrong, we could not figure out where we were. Four times we passed by ancient hollow cypress, but could not find a passage through this creek. We paddled misdirected for hours and never saw another face. It rained cold heavy drops on our unprepared heads. Now the larger river we were on narrowed to sandy sweet grass channels and the old Bartram canoe trial signs reappeared in unlikely places. Often behind curtains of Spanish moss. We probed the narrow sweet grass channel, which turned into a high banked creek littered with fallen wood. Mist gathered and daylight evaporated as we were forced into creative and highly physical maneuvering through the water. We had now gained a bit of skill in hauling over logs or gathering momentum to break through debris dams. As we unpacked our flashlights the creek entered a big water river. What to do now? It was night and we could see little ahead. We were wet from head to toe as the night chill was setting in. I wanted to go ashore and construct a lean to and fire, while Christy wanted to push ahead, all night if needed. Once again, we had no idea where we were, but we paddled down river half a mile and saw a light. It was the only light, for even the stars were masked. We raced toward it and met, yet again, two sportsmen, hunters this time, all bedecked in camo. They informed us that we were on the Tensaw, but not the right Tensaw. We had pushed much too far west and were now on the western Tensaw River. Bad news. These fine gentlemen were returning from a weekend of duck hunting and they offered us to sleep in their houseboat. They had even left a fire. We thanked them and they headed off while we paddled to their rustic retreat. Our impression of the outdoor enthusiasts of Alabama continued to favorably expand. We built up the dying embers of their fire while drying our clothes and eating bags of chips. The cabin, though amply populated with mosquitoes, provided a welcome sanctuary from the nights winds and rains.

"Fortunately the wind was at our backs and so after heaping bowls of apple flavored grits we piled into the canoe, opening Dana's parasol for an extra push in the sharp breeze. We located our position on the map and concluded that we were probably more than a day's paddle from the landing, but we remained uncertain of our precise position. A mile down river we spotted a small cabin with a boat tied to the dock. Christy rushed up to the cabin with our blob of a map to confirm our position. For some time she did not return, and Dana and I were convinced that she had stumbled upon two hunters in a loving embrace and they had been forced to kill her. However, in fact, she had woken up a pair of hard drinking sportsmen who proved to be courteous, generous and remarkably helpful. Buddy and Cris, two outdoor-loving, native Alabamians, confirmed our fears that we remained more than a day away from Hubbard's Landing. However, they were packing out and would happily tow us back to their landing site and Cris offered to drive us back to our awaiting cars. They popped a beer, vacuumed the floor, cleaned the dishes and loaded the boat while we shivered around the kerosene heater. Here was our rescue team. These guys saved our asses and Christy's job, and shared their good stories. As I rode in the towed canoe our friends chain smoked, guzzled beer and entertained Christy and Dana with their very best snake and alligator tales. We owe a hearty thanks to three pairs of Alabama's finest outdoor sportsmen."

Breakfast of champions — the expeditions team is saved by the fine sportsman of Alabama.

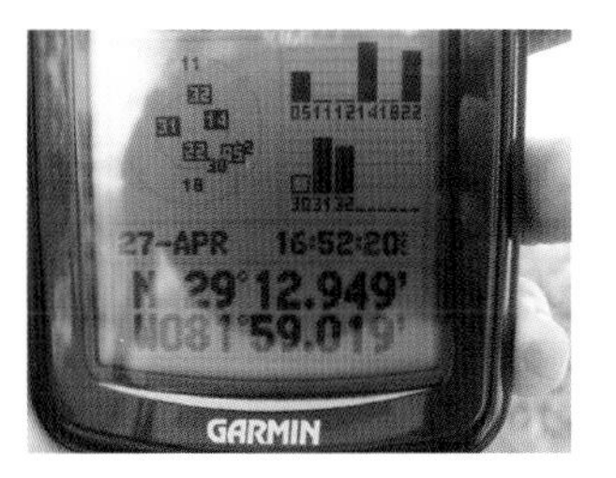

Each of us experienced stress dreams and mosquito bites during the course of the night. In the early hours before dawn the wind raged outside and we realized that if the wind continued at this velocity we would not be able to canoe out of here.

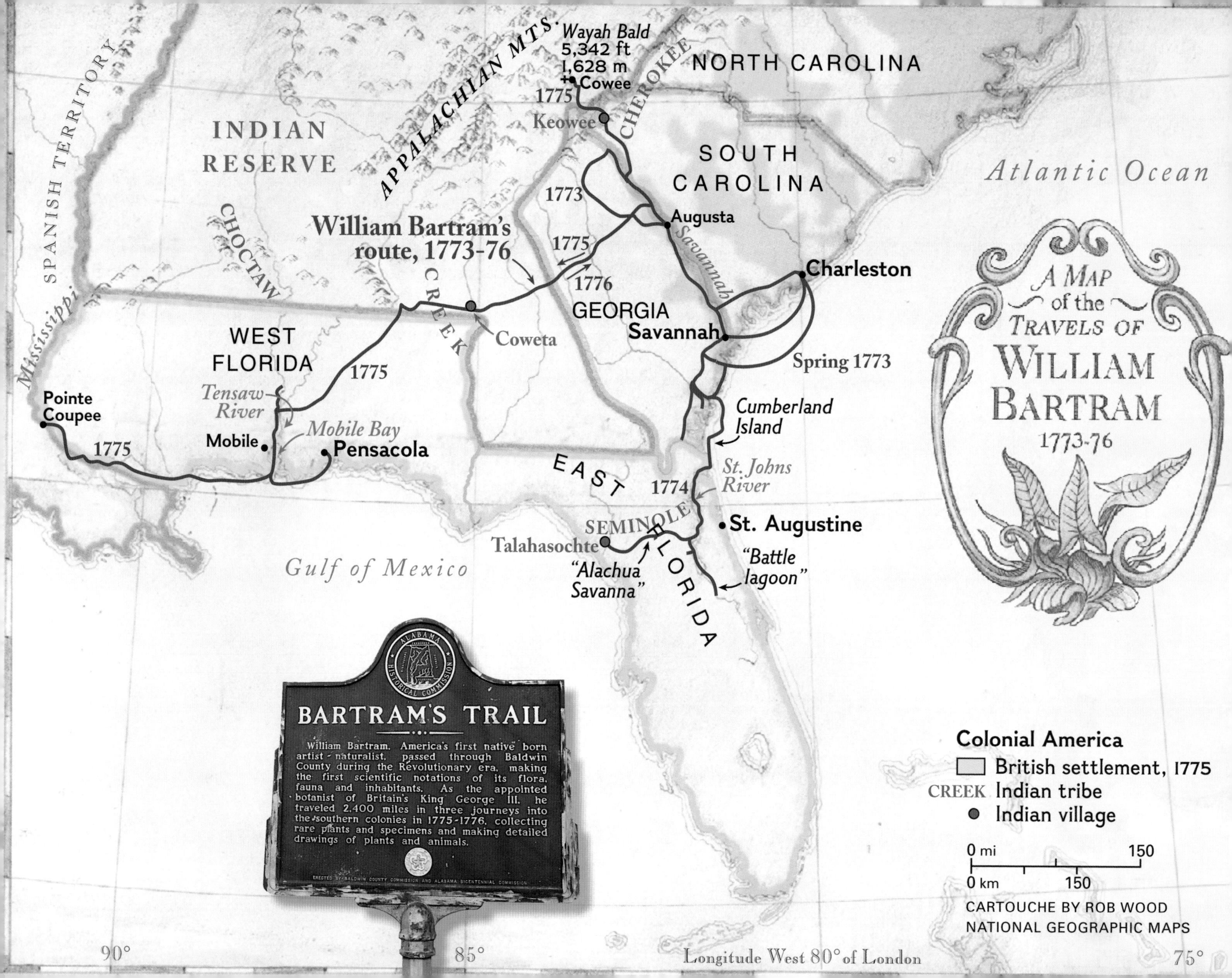
A Map of the Travels of William Bartram 1773-76
William Bartram's route, 1773-76
Wayah Bald
5,342 ft
1,628 m
Cowee
Keowee
APPALACHIAN MTS.
CHEROKEE
NORTH CAROLINA
SOUTH CAROLINA
INDIAN RESERVE
SPANISH TERRITORY
CHOCTAW
CREEK
Mississippi
Atlantic Ocean
Augusta
Savannah
Charleston
Spring 1773
GEORGIA
Savannah
Coweta
WEST FLORIDA
Pointe Coupee
Tensaw River
Mobile Bay
Mobile
Pensacola
EAST FLORIDA
SEMINOLE
Cumberland Island
St. Johns River
St. Augustine
Talahasochte
"Alachua Savanna"
"Battle lagoon"
Gulf of Mexico
1773
1774
1775
1776
Colonial America
British settlement, 1775
CREEK Indian tribe
Indian village
0 mi 150
0 km 150
CARTOUCHE BY ROB WOOD
NATIONAL GEOGRAPHIC MAPS
90°
85°
Longitude West 80° of London
75°
ALABAMA HISTORICAL COMMISSION
BARTRAM'S TRAIL
William Bartram, America's first native born artist-naturalist, passed through Baldwin County during the Revolutionary era, making the first scientific notations of its flora, fauna and inhabitants. As the appointed botanist of Britain's King George III, he traveled 2,400 miles in three journeys into the southern colonies in 1775-1776, collecting rare plants and specimens and making detailed drawings of plants and animals.
ERECTED BY BALDWIN COUNTY COMMISSION AND ALABAMA BICENTENNIAL COMMISSION

DEEP SOUTH REGION
WILLIAM BARTRAM TRAIL
TRACED 1773-1777
Eminent artist - naturalist. Described numerous species of flora including Franklinia. Explored local area in 1773.
ERECTED BY
The State Botanical Garden of Georgia
IN COOPERATION WITH
The Garden Club of Georgia, Inc.
DEEP SOUTH REGION
WILLIAM BARTRAM TRAIL
TRACED 1773-1777
In 1775 William Bartram described Pensacola and the hospitality he received when even Gov. Chester urged him to be his houseguest.
ERECTED BY
Althea Garden Club
IN COOPERATION WITH
The City of Pensacola, The Architectural Review Board, Florida Federation of Garden Clubs, Inc.
William Bartram Memorial Park
THE STATE OF FLORIDA
DEPARTMENT OF NATURAL RESOURCES
FLORIDA RECREATION DEVELOPMENT ASSISTANCE PROGRA
DEPARTMENT OF STATE
HISTORIC PENSACOLA PRESERVATION BOARD
THE CITY OF PENSACOLA
PARKS AND RECREATION DEPARTMENT
DEPARTMENT OF ENGINEERING AND FIELD OPERATIONS
THE MOUNT ROYAL SITE
Mount Royal has been a favored location for people to live for thousands of years. Archaeological sites include a Native American burial mound, earthworks, village area, and evidence of a British plantation, as well as the remains of a Spanish mission occupied by the Timucua Indians. British naturalist William Bartram visited Mount Royal in 1765-1766, and again in 1774. His description of the large mound, fields, earthen causeways and an artificial pond was published in 1791 and is one of the earliest accounts of an Indian mound in North America. Bartram's plan of the mound was later published in 1848 by the newly formed Smithsonian Institution. Archaeologist Clarence B. Moore excavated the mounds in 1893 and 1894. Moore found human burials with hammered and embossed sheet copper ornaments, polished stone tools, pearl and shell beads, and decorated ceramic vessels. The copper ornaments are similar to those found at Mississippian sites in Georgia, Alabama, and Oklahoma and date between 1000 and 1500 A.D. Archaeologist B. Calvin Jones' salvage excavations at the village site in 1983 and in 1994-1995, revealed evidence of six structures. These buildings contained Spanish artifacts and were probably part of the Mission of San Antonio de Anacape (1587-1675).
F-411
FLORIDA HERITAGE LANDMARK
THE FLORIDA DEPARTMENT OF STATE
1999

April 21, 2008

"Fleabag: Our trip to Flea-world, "America's Largest Fleamarket" was not quite what we expected and we came home pretty much empty handed, except for a dozen or so cheap machetes that Mark picked up. What does he do with them all? It was hard to resist the market's tasty morsels on offer, but we couldn't having stopped for some pretty good southern bar-b-que on the way."

Fully Flexed Burial

IDA MAE & JOE'S
NORTH MIDWAY
RESTAURANT
Waypoint
CHARALSTON
17-NOV-07 15:57
Location
N 32°46.826'
W079°55.878'
Elevation
Depth
32
Show Name on Maps
Goto
Map
OK
BIG CAT
ENCOUNTER
12, 2 & 4pm • FRI - SAT - SUN • APRIL 4 - 27
START YOUR BUSINESS
AT FLEA WORLD
START AT $25 WEEKEND. 50-60 THOUSAND CUSTOMERS EVERY
2
3
4
5
6
7
THIS TABLE
$5.00
EACH
WM. BARTRAM
MOUNT ROYAL Ave.
OPEN 7AM - 9PM
DAILY SPECIALS
SHRIMP
FISH CRAB CAKES

"I only wish I could stay awake late enough to enjoy the majority of it."

"Big Top", New Orleans

"We've just left the shores of lake Pontchartrain, which Bartram sailed across in 1775. Unlike Bartram however, we detoured south to the metropolis of New Orleans where we spent the Thanksgiving holiday feasting on smoked, deep-fried turkey with Dana's lively comrades: Courtney, Raven, Jayme, Jackie, Mike and others as well as assorted dogs. We were guests of Raven and Jayme in their comfortable double shotgun house in the Bywater district. Here we spent days richly entertained in between collection excursions. The wild lands and parks we visited were either closed to access or deeply scarred by Hurricane Katrina.

"Dana associates with a cultural community here which shares an affinity with the aesthetics with the circus, carnival and vaudeville tradition. This creative interrogation of the carnivalesque doubtlessly owes a debt to the cultural primacy of Mardi Gras, but also extends to include popular entertainments as diverse as the high wire, burlesque and even puppet theater. Not only are our friends passionate about getting dressed up, but they are also accomplished musicians and performers. The music, brilliantly composed and performed, references numerous stage, parade and big-top traditions is diverse, generous and distinctively bawdy. In fact, all of these expressive forms flirt with the seedier side of adult entertainment. They also consciously exploit the political potential of the carnivalesque, which employs parody and satire in not too subtle doses. For example, one night we attended a violent and hysterical puppet show, which took the form of a game show, and although conducted entirely in gibberish its insightful critique was entirely accessible and highly amusing. I'm impressed with the vibrance of the cultural scene, particularly with regard to music and performance. I only wish I could stay awake late enough to enjoy the majority of it."

above, *Courtney Lain and the Herringbone Orchestra at the "Big Top.", New Orleans;* right, *Hurricane ravished Fontainebleau State Park.*
previous spread, 1-, 6-, 7- Mount Royal, FL, 2- Midway, GA, 3- Charleston, SC, 4- Flea World, FL, 5- Deland, FL,
following spread, 1- *Scott Flea Market, FL,* 2- *Mark's equipment, Philadelphia, PA,* 3- *Hindi shrine, Disney World, FL,* 4- *M.K. Rawling's home, Alachua Co., FL,* 5- *Atlantic Center for the Arts exhibition, FL, October 10, 2008*
pages 64-65, 1- *Black-Crowned Night Heron* 2- *Cuban Anole* 3- *Channel Cat* 4- *Limpkin* 5- *Florida Brown Snake* 6- *Wildhorse* 7- *Banded Tussock Moth caterpillar* 8- *Silverback Gorilla* 9- *Tree Snail* 10- *Apple Snail Eggs* 11- *Black-Crowned Night Heron* 12- *Sheep's Head*

1 2
3 4
Del Monte
BANANAS
INSECT PINS
ENTO PINS
marjorie kinnan rawlings
v pomladi življenja I

Eastern Birds
Seashore
FISHES OF THE
The Wild Edge
An Outdoor Guide to Bartram's Travels
Travels of William Bartram
Post Cards
STAEDTLER

Details of grave markers, Savannah, GA.

JAMES T. WELSMAN,
SON OF
JAMES & AMELIA WELSMAN
BORN AUGUST 7, 1813;
DIED JULY 7, 1877.
"THE UPRIGHT SHALL DWELL IN THY
SACRED to the MEMORY
GRACIE

WILLIAM BARTRAM'S
OF PAYNES
APRIL, 17
ON A SUDDEN OPENS TO
1
2
7
8
13
14

4
5
6
10
11
12
16
17
18

"Behold him rushing forth from the flags and reeds. His enormous body swells. His plaited tail brandished high, floats upon the lake. The waters like a cataract descend from his opening jaws. Clouds of smoke issue from his dilated nostrils. The earth trembles with his thunder."

— William Bartram

Mark Dion's
Travels of William Bartram - Reconsidered
Gregory Volk

1. A Cabinet of Alligators

While addressing the large trajectory of Mark Dion's *Travels of William Bartram—Reconsidered,* his partly earnest, partly hilarious, sharply insightful, intelligent, and at times eccentric retracing of Bartram's 1773–77 knowledge-seeking quest through the American Southeast, I'd like to look closely at the actual exhibition resulting from this excursion. Bartram's Garden, in Philadelphia, offers valuable insight into how John Bartram, who is often called America's first botanist, and his artist/writer/scientist son William, often called America's first naturalist and travel writer, lived, worked, and thought. As you gaze at 18th-century furniture and architecture in the pristine and well-preserved Bartram family house, you also see botanical specimens, drawings, and scientific tools that indicate how advanced science was practiced in the era, and when you go outside into the gardens you begin to realize how much this Quaker place was a confluence of science, art, religiosity, and nature. Dion's exhibition, interspersed in rooms around the house, largely consists of period and specially designed new cabinets filled with sundry objects he collected during his travels, including flora, fauna, and various cultural items, melds with this milieu (to the point where it is sometimes tough to tell what is from the Bartrams and from him, what is old and what is new) but also takes things in startling new directions. For one thing, Dion's flora and fauna include such unremarkable stuff as seashells, pods, leaves, bark, acorns, sea wrack, and a dead snake lolling in a jar of alcohol, while his cultural artifacts include rusty nails, a rubber stamp, old tools, and recent bottle caps, all of which scramble distinctions between high and low, rare and common, valuable and valueless. What happens is a visual and ideational dialogue between Dion and the Bartrams as kindred, questing spirits; between nature, its representations, and systems of classification now and as they were two hundred years or so ago; between the 21st and 18th centuries, and, really, between the United States at its origins and what the country has become. In effect, Dion has acted as both a quirky explorer and an anthropologist, gathering a crazy assortment of eclectic stuff which he uses to examine not a strange, exotic civilization centuries ago, with curious customs and beliefs, but instead his own American society right now; also not the exotic wilderness of William Bartram's day but instead current nature, crisscrossed by interstates, truncated by suburban and exurban sprawl, top-heavy with the tourist industry, jam-packed with box stores, and altogether molded by its human inhabitants.

I'd like to look closely, for instance, at Dion's alligator cabinet, displayed in a tall old wood cabinet with mullioned glass front, and fitting easily, yet on closer inspection jarringly, with the other displays and memorabilia in the Bartram family house. While traveling in Florida (populated by alligators basically since the extinction of the dinosaurs) William Bartram encountered alligators in the wild, and at peril to himself. Here is how he describes one such encounter: "Behold him rushing forth from the flags and reeds. His enormous body swells. His plaited tail brandished high, floats upon the lake. The waters like a cataract descend from his opening jaws. Clouds of smoke issue from his dilated nostrils. The earth trembles with his thunder." Traveling in the same region, Dion interacted with and collected not alligators at all, but alligator-themed knickknacks, kitchen items, household goods, and souvenirs, which are displayed on shelves in the cabinet. A famous reptile—after all, Florida's Official State Reptile and an enduring symbol of the state—has been sent through the wringer of popular culture, entertainment culture, tourism, marketing campaigns, sentimentality, and commerce, and the transmuted creatures that result are real nature/culture collisions: part schlock; part curios; part practical stuff with a reptilian twist; part peculiar, latter-day versions of ancient animal deities. While investigating nature, just as William Bartram did long ago, Dion focuses on the many ways that nature has been (and still is) distorted,

In Midway Georgia where Bartram visited in April, 1773.

left, *Alligator of St Johns, from the botanical and zoological drawings (1756-1788) by William Bartram.*

I took the team past the richly planted central access exhibits and straight to do what one always does in Disney's realm: wait in a long line. We waited for about 45 minutes for the Jungle African Safari. The associates were excited and eager and this respite gave us time to catch up. Soon we were bumping about in a vast ride watching the famous animals of the African Continent. For Nancy, Julie and Bronwyn it was quite hard to tell what was real and what was artificial. They imagined the crocodiles

(continued on page 71)

transformed, and endlessly mediated by us: the million ways that we project our ourselves and our ideologies on the natural world.

Dion's alligator collection begins at the upper left with two coffee cups. The first, which succinctly sports the word "Florida" on its side, has an alligator with an open mouth protruding from the side, while its tail doubles as the cup's handle. The fierce, wild, living thing, which you fear might gnash you in its big jaws if it had the chance, has now been tamed, domesticated, diminished, and surreally transformed into a humorous, yet utilitarian, object guaranteed to make you crack a smile. The second, green coffee cup is an even more interesting specimen, for it features a frolicking alligator that looks very like a beagle, and a cute, seemingly smiling, baby alligator climbing just above the rim. One alligator has been transformed, willy-nilly, into a domesticated doggie, the trusty and companionable pet who bounds around your yard or living room, and therefore is no cause for worry anymore, suggesting that some animals (like beagles) are "good," while others, like alligators, are very bad and threatening, and pay no attention, of course, to the fact that alligators are much more threatened by us than vice versa. The other has the fetching sweetness of a toddler playing peek-a-boo, perhaps with residual memories of "Kilroy was here" doodles made by U.S. servicemen during World War II and the Korean War.

Throughout the collection, alligators are typically cute, and they are relentlessly playful, happy-go-lucky, sometimes cuddly, and altogether positive—in a nation famous for its optimism, a country (as the politicians like to say) whose best days are always, always still to come. One finds a clickable alligator; an alligator with an attached pump that you can use to make it move; a bizarre pen whose body is an alligator; a pink, blue, and yellow beaded alligator that is a key chain; a green alligator ready to be welcoming kitsch on someone's doorstep; two sweetie-pie ceramic alligators with open mouths and pink tongues that seem to have anticipated "Barney" by several decades.

A carnivorous reptile that has perfectly adapted to its surroundings over hundreds of millions of years becomes, in effect, "our" pet and mascot, as well as a symbol of our imagined mastery of the natural world, and it does so with gusto and variety, evoking folk art, cartoons, cheap prizes at the carnival, Surrealist imagery, mass-market knockoffs of Native American totems, and advertisement logos. In a water globe, two green plastic alligators lounging under a palm tree appear to be smiling or laughing, one of many times when these reptiles are invested with cheery, breezy, sunny dispositions. "Don't worry," these things seemingly announce, "Buck up, keep your chin up, keep on the sunny side, welcome to paradise. Have a nice day." Yet most of these specimens are old, sullied, used, and often damaged. These are souvenirs with a past, uprooted flotsam with a disconcerting aura of loss, dislocation, and, implicitly, mortality, for people owned these knickknacks once, but those people may very well have vanished from the earth. They are not precious relics from centuries ago, but instead nonprecious leftovers from a few decades ago, and they evoke a bygone era, a prior Florida steeped in fantasies of relaxation, playfulness, and youth, a more innocent Florida of the 1930s, '40s, and '50s which is anachronistic and is altogether fading from view, especially now when things seem anything but carefree. That's when Dion's antic humor takes on a darkly serious side. Chuckle-inducing tchochkes are one thing. Aged tchochkes that have survived their vanished owners, and that make you think of our fragility and brevity, and of how our temporary lives (and temporary civilizations, for that matter, no matter how powerful and prideful), are soluble in nature and dominated by the vast scale of time are something else entirely, and considerably unnerving.

Of course there is something ironic in Dion's alligator collection. Of course there is something ridiculous and laughable about undertaking an epic voyage of discovery, only

FLORIDA

ALLIGATOR WRESTLING
Alligator, Parrot Jungle, Miami, Florida
Breeding Pool at the Ostrich-Alligator Farm, St. Augustine, Florida
The Oldest City in the United States
Hungry Alligators at Tropical Hobbyland, Miami, Florida
Alligator, Parrot Jungle, Miami, Florida
ALLIGATOR HUNT IN THE EVERGLADES FLORIDA
N-48
PHOTO BY J.J. STEINMETZ
Spectacled Caimans, Philadelphia Zoo
SK5227
Alligator Eggs Hatching, Florida.
MUSA ISLE SEMINOLE INDIAN VILLAGE
ALLIGATOR WRESTLING
Hungry Alligators at Tropical Hobbyland, Miami, Florida
Breeding Pool at the Ostrich-Alligator Farm, St. Augustine, Florida

to acquire a sprawling assortment of cheesy alligator doodads in flea markets and antique shops, and of course there is something downright goofy about seeing such doodads treated as important specimens fit for extensive contemplation and analysis. Still, Dion's alligator collection is curiously evocative, probing our aspirations and consternation, as well as our attraction to, alienation from, confusion about, and endless urge to dominate the nature which surrounds us (and of which we are a part).

We are often oblivious to and not that knowledgeable about the animals with which we share an ecosystem, possibly because we're not all that great about the concept of sharing an ecosystem altogether. At the time of this writing, fervent delegates to the 2008 Republican convention in Minneapolis rocked the arena with chants of "Drill, baby, drill," calling for more oil and fossil fuel consumption in a time when global warming is increasingly perceived as an imminent and perhaps catastrophic threat, and not just to us, but pretty much to life everywhere. Several mutant alligators collected by Dion seem spawned by our confusion and scant knowledge. They conflate mammals and reptiles, wildly divergent ecosystems, North America and places far away, like Africa or Australia: a silver one, for instance, with an upraised snout that looks for all the world like a hippo, and another befuddling one on an ashtray that's like a cross between a gator, a platypus, and a seal. A shiny metal alligator stands on a chest or box, which is matter-of-factly labeled as a "Souvenir of Florida." On one level we are a practical, can-do nation and here is our practicality in droves, for you just can't get much more plainspoken and straightforward than a "Souvenir of Florida." We are also a nation enthralled with entertainment, and a paperweight in the form of a good-humored, yet sly and knowing gator conjures a singer up on the stage in some smoky nightclub of the past, a sultry Marlene Dietrich wannabe in spiky scales instead of a sequin dress, or a rakish, Dean Martinesque crooner with a good heart and leering eyes—it is impossible to tell the alligator's gender. In a nation whose utopian and paradisal inclinations stretch all the way back to the Puritans and their great desire to erect a "New Jerusalem" on the Massachusetts coast— this "quiet, pleasant, greening land called America" as Ronald Reagan put it in 1988—old picture postcards of tourist attractions (Alligator Alley), motels (Gator Lodge, in Jacksonville), and alligators in zoos (including a risqué gator poised to bite a perky woman in the rear end) have the look of frayed, decaying idylls. Several ashtrays hark back to a time when smoking was commonplace, an increasingly distant time of highballs and Lucky Strikes on the veranda or beside the lovely pool, and therefore they are cultural dinosaurs. In one, a brown alligator with an enthusiastic open mouth lounges on the bank of a swamp, with the swamp being the bowl where stubbed out butts, ash, and copious tar would form a noxious gruel. Possible, no doubt unintended, metaphors abound, for instance alligators trying to survive in polluted waters, like the Everglades ravaged by phosphates from the sugar cane industry or Silver Springs despoiled by nitrates. Then again, Americans have long had a reverence for wilderness, even as we excel at damaging and eradicating that which we allegedly adore.

(continued from page 68)

as fake while they are flesh and blood and understood the baobab trees as real when they are not. All too soon it was over and we were on to hike the gorilla trail which they rushed through like clockwork dolls.

Before long we were on another line threading through the crazy entrance to the ride. First you go through a run-down Third World hotel lobby, past a shrine in which the Hindi gods have been transformed into Bigfoot, past an expedition outfitter and lastly through a funky Yeti museum.

Type Specimen
Franklinia alatamaha Bartram
TRAVELS OF
William Bartram
RECONSIDERED
Seed Collection
M.Dion 2007-2008

2. A Cabinet of Plants, a Cabinet of Water

William Bartram certainly has his champions and aficionados, but I suspect that Mark Dion's exhibition will inspire many people to read Bartram's great 1791 book, *Travels through North and South Carolina, Georgia, East and West Florida...* known more simply as *The Travels of William Bartram,* for the first time. When they do, they will find that it is a marvelous mix of elegant prose, close scientific scrutiny, astute cultural inquiry (including the uneasy relationship between encroaching white traders and settlers and Native American tribes), and convincing spirituality. In studying and chronicling the natural world, Bartram was not only engaging in serious scientific research, but was also, in his opinion, directly experiencing God's spectacular creation. He was witnessing thrilling evidence of the Maker's hand, and there are many times in his book when copious scientific observation concludes with religious praise and exhortation.

Bartram's book, in which a protagonist leaves a cultured, civilized place (Philadelphia) to explore a far off, exotic region is a fascinating extension of the first great travel narrative produced in America after European settlement: William Bradford's *History Of Plymouth Plantation*. It also anticipated any number of stellar travel narratives coming much later, including Herman Melville's *Moby Dick*; parts of Walt Whitman's *Song of Myself*; Frederic E. Church's paintings of the Andes; paintings of the American West by Church, Albert Bierstadt, and Thomas Moran; Jack Kerouac's *On the Road*, and countless road movies, among many, many others (Mark Dion's road trip, chronicled in an exhibition, website, photographs, and journal entries thus takes its place among a type of literature and art that has proved crucial in America). Bartram's spiritually charged approach to nature strongly influenced the English Romantic poets Samuel Taylor Coleridge and William Wordsworth, and also anticipated the Transcendentalist poet/philosopher Ralph Waldo Emerson, in Concord, Massachusetts, who counseled immersive and ecstatic encounters with nature, understood to be suffused with divinity and eternal spiritual truths, which could then be channeled into poetry and art.

Although Emerson, who essentially shifted his spiritual locus from organized religion to nature, theorized such ecstatic encounters, he rarely attempted to describe what they actually felt like. There is, however, a famous passage in his essay *Nature* that does so wonderfully. The passage begins, "Crossing a bare common, in snow puddles, at twilight, under a clouded sky, without having in my thoughts any occurrence of special good fortune, I have enjoyed a perfect exhilaration. I am glad to the brink of fear." Emerson then goes on to memorably declare, "I become a transparent eye-ball; I am nothing; I see all; the currents of the Universal Being circulate through me; I am part and parcel of God."

Becoming a "transparent eyeball"—with that kind of radical openness to and fluid exchange with the world, really with an almost egoless merging of self with world—is exactly the kind of thing that inspired and energized artists and writers in Emerson's day, but also many others coming much later. The Hudson River School painters were indebted, as were Luminist painters like Fitz Henry Lane and Martin Johnson Heade, whose shimmering seascapes and landscapes involve detailed depiction of the physical world, but also seem reverential, ethereal, meditative, and imbued with pure spirit. Emerson's friend and neighbor Henry David Thoreau, the author of *Walden*, was likewise inspired (the naturalistically inclined Thoreau, who authored several nature excursions, also seems extremely close to William Bartram), as was Walt Whitman, who once famously declared, "I was simmering, simmering, simmering; Emerson brought me to a boil." Many subsequent artists and writers share core-level Emersonian inclinations, including Edward Hopper (who read Emerson extensively), Barnett Newman (in 1936, the newlyweds Barnett and Annalee Newman traveled to Concord on their honeymoon, in order to visit sites associated with Emerson and Transcendentalism), Agnes Martin,

We were kept quite busy this afternoon since horticulturist, Leslie Reed, had organized a plant collecting project which had a Bartram inspiration. There is a plant which Bartram called Smilax, or pseudo china-berry, a type of catbrier which can be used as food and had been by the native americans. It is a leaf of three heavy lobes and a woody vine. They dug some impressive tubers and I washed and cut them into small pieces while Leslie mashed and boiled them. A lovely amber colored starch was produced, but it tasted bitter. Bartram said it was dried to a powder and mixed with corn meal and fried in bear fat.

left, background, *Franklinia alatamaha, Franklin tree, specimen by William Bartram, 1773-76.*

CAROLI LINNÆI

Medic. & Botanic. in Acad. Upsaliensi Profess.

Reg. & Ord.

GENERA PLANTARUM

Eorumque

CHARACTERES NATURALES

SI

Apud { CONRADUM WISHOFF, ET GEORG. JAC. WISHOFF, Fil. Conr. } 1742.

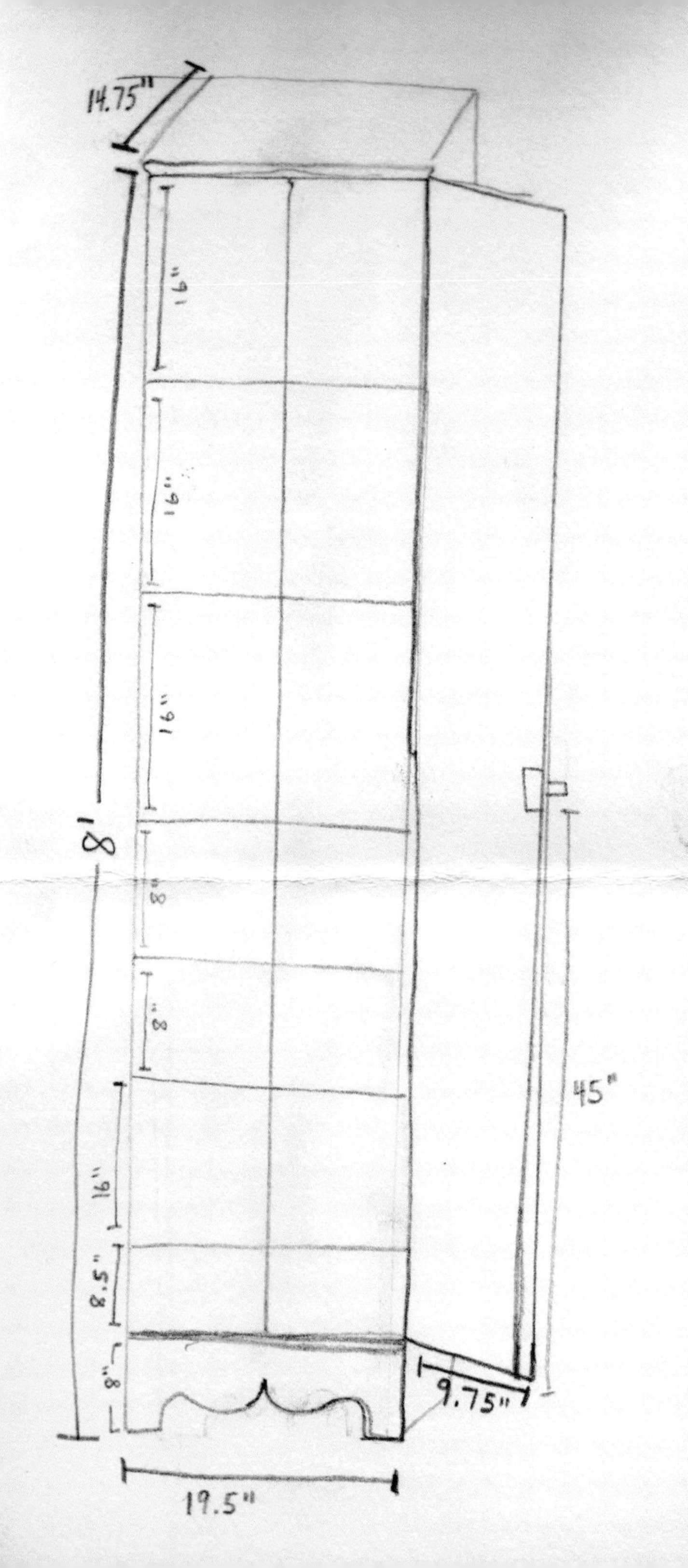

About Linnaeus

Carl Linnaeus was one of the most influential scientists of his time. Determined and convinced of his own abilities, he made it his life's work to develop and refine a way to classify and name all life on Earth. So straightforward was his new naming system, it is still used by scientists today.

Born in southern Sweden in 1707, Linnaeus studied medicine at the Universities of Lund and Uppsala in Sweden. Because most medical treatments in those days used products derived from plants, it meant he also studied botany, a childhood fascination for Linnaeus. It was while studying in Sweden that Linnaeus prepared one of his most influential works, Systema Naturae (1735). A slim but important volume, in it he proposed a new way to classify, or group, the natural world. He started by dividing nature into groups based on shared physical characteristics. First came three kingdoms: plants, animals and minerals. Kingdoms were divided into classes and then into orders, which were further divided into genera (singular: genus) and then species (singular: species). Rather shockingly for the time, Linnaeus grouped humans along with the primates, and also divided the plants by the number and arrangement of their reproductive parts. Despite blushes in polite society, scientists quickly accepted his sexual system for plants. Over the years, Linnaeus tirelessly added detail to the original 11-page work until, by the tenth edition, it was a substantial two-volume publication of 1,384 pages. It is from this book that the title of the exhibition, *Systema Metropolis*, is derived.

But his work didn't stop there. Linnaeus took on the challenge of finding a simple method of naming species. In the early eighteenth century, scientific names for species were already in Latin, but were often long and unwieldy. For example, the humble tomato was a bit of a mouthful: *Solanum caule inermi herbaceo, foliis pinnatis incisis, racemis simplicibus*. Starting with his monumental work *Species Plantarum* (1753), Linnaeus gave all the plants then known a simpler Latin name in two parts, known as a binomial. The first part was the genus, followed by the species. Using this system, the tomato became a more palatable *Solanum lycopersicum*. He gave binomial names to animals five years later and, between 1753 and his death, named thousands of plants and animals in this way. This binomial system was adopted by other scientists and became the standard way of naming organisms.

Throughout his life Linnaeus worked to understand the natural world and the systems that govern it. He wrote not just about classification but also ecology – how organisms interact with their environment. He described food chains and even explored the concept of race, suggesting humans could be divided into at least four groups: Americanus, Asiaticus, Africanus and Europeanus. Forced to retire from teaching in 1774 by ill health, Linnaeus suffered a series of strokes and eventually died in 1778. He was driven by a love for the natural world, and the desire to understand and name it. His legacy remains, and is used by the many dedicated scientists today driven by that same desire.

— *2008, the Natural History Museum, London*

far left, *title page from John and William Bartram's copy of Linnaeus's 1742 Genera Plantarum;*
left center, *Sherwood's drawing with measurements taken from Linnaeus's herbarium cabinet;*
above, *Detail from Dion's exhibition "Systema Metropolis" at London's Natural History Museum, 2007;*
right, *Dion's herbarium cabinet, 2008*

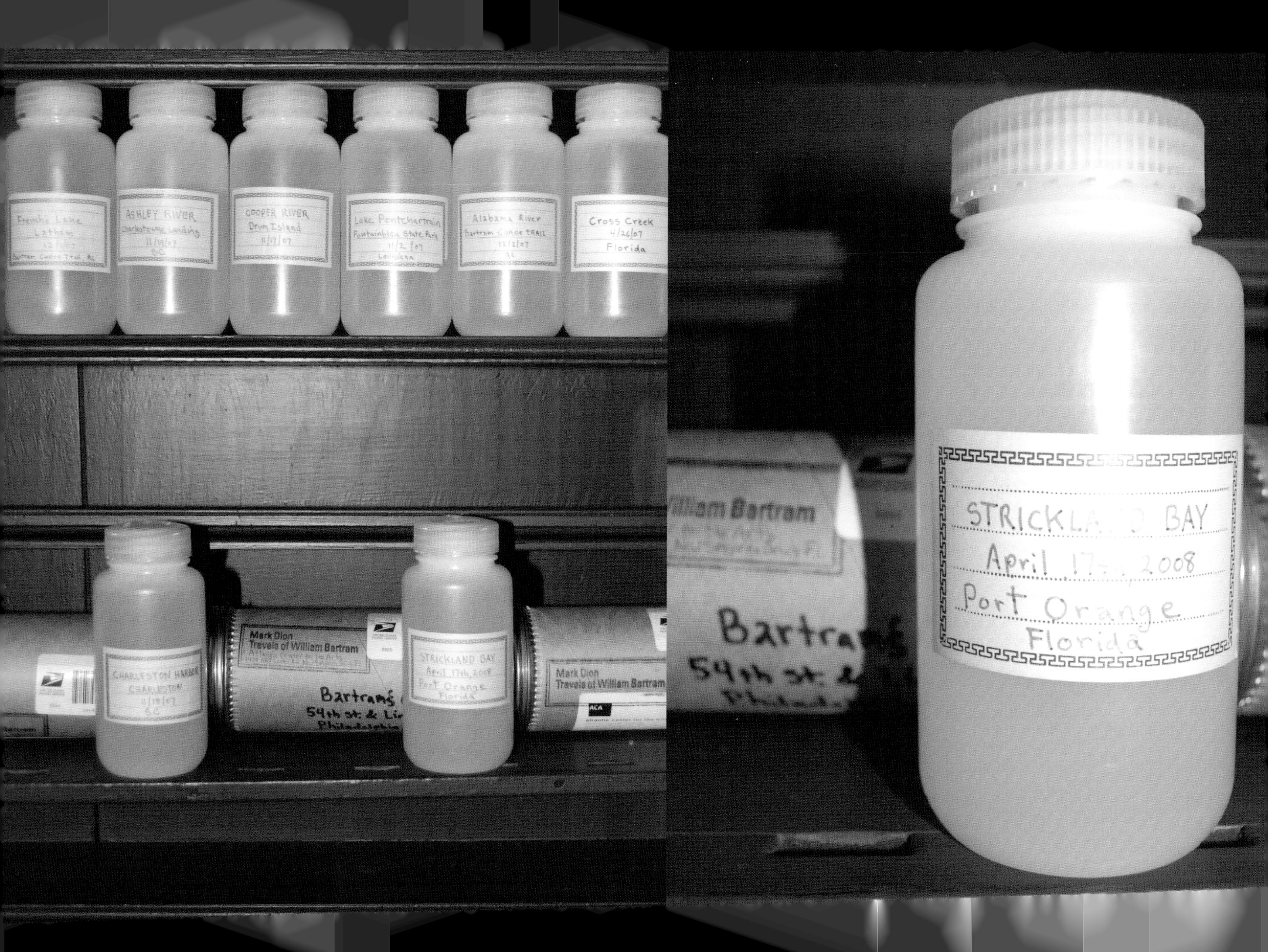
French's Lake
Latham
ASHLEY RIVER
Charlestowne Landing
SC
COOPER RIVER
Drum Island
Lake Pontchartrain
Fontainblea State Park
11/2/07
Louisiana
Alabama River
Bartram Canoe TRAIL
12/2/07
AL
Cross Creek
4/26/07
Florida
CHARLESTON HARBOR
CHARLESTON
SC
Mark Dion
Travels of William Bartram
STRICKLAND BAY
April 17th, 2008
Port Orange
Florida
Mark Dion
Travels of William Bartram
William Bartram

with her sublime, abstracted landscapes, and Robert Smithson with his earthworks, notably *Spiral Jetty*. When you actually visit *Spiral Jetty*, and spend considerable time there, a couple of important things become apparent. For one, the work does not seem very monumental, as it often does in photographs. Instead, it seems rather modest, even humble: not an earthwork imposed on the landscape but rather one that merges and blends with its surroundings. It also constantly changes over the course of a day, from bedazzling white in pinkish red water to a somber grayish black in blue water. Even though it is largely made of stones, it seems hyper alert to everything in its vicinity: to the vicissitudes of changing light, passing clouds, the great horizontal expanse of the lake, and really to a vast scale of time, including cycles of creation and destruction. The work also has a tremendous impact on one humanly. You feel contemplative and concentrated but also remarkably open and invigorated.

With Emerson's "transparent eyeball" in mind, consider Mark Dion's cabinet of plants. In a narrow cabinet specially designed for this exhibition, manila folders tied with blue or gold string are filled with botanical specimens acquired during the artist's travels, which are lightly attached to sheets of paper. Several unbound stacks allow one to hold and closely examine individual items. As you scout through these stacks, you see dried ferns, leaves, stalks, and slight branches which seem like the specimens in a botanical collection, perhaps even an 18th-century one, but nothing is labeled and no information is provided. It is impossible to know if these are precious or commonplace specimens, where they are from, what they signify, what their dates are. This is a classification system that offers no knowledge per se, and that basically prevents one from fitting things into any category, but then again it inspires one to look freshly, even with wonder and awe, at nature, or rather fragments of nature. As you look at these specimens, they suggest intricate drawings perhaps made by Dion; elegant Chinese landscapes, say by Song-dynasty painters of the 12th or 13th century such as Li Tang and Xia Gui; and William Bartram's excellent botanical drawings, some of which are displayed in the Bartram house—except for the fact that nature itself becomes the art: you're not looking at an artist's drawing of a leaf, but instead at an actual leaf which doubles as a drawing. Moreover, a single, dried, fragile leaf nestled on a white expanse of paper or a gracefully curving fern frond are gorgeous and enthralling, and have a rapturous effect, almost as if you're in the presence of something sacred and sublime, a beauty capable of inducing rapt contemplation and exultation, and that's prime Emersonian territory. For me, this is one of the keys to Mark Dion's work altogether. For all his interest in science, natural history, and collecting, and for all the savvy ways he questions how knowledge of the natural world is displayed in museums, he's one of the excellent practitioners of a contemporary sublime, in which encounters with nature, however mediated, are still cathartic, and still have a mind-bending, soul-expanding power.

Consider Dion's cabinet of water samples as well. From time to time while traveling, he collected samples from lakes, rivers, springs, and the ocean in uniform plastic bottles, which he then sent to Philadelphia in mailing tubes. Normally, such samples would be used for something specific, to measure the water for purity or pollution, for example, but no such measurements were taken. Instead, bottles filled with water are displayed on shelves in a pre-existing cabinet in the study, along with the mailing tubes, and labels announcing the date and place where the samples were taken, like Strickland Bay in Port Orange, Florida; Lake Pontchartrain in Louisiana, the Alabama River, in Alabama, and Manatee Springs in Florida. Dion's varied arrangement of these bottles and mailing tubes on six shelves is very sculptural, suggesting the rudimentary geometric forms, simple mathematical progressions, and serial repetition of Minimalist sculptors such as Donald Judd and Carl Andre. It likewise suggests an admixture of a scientific laboratory and designer water for sale in an upscale shop. Instead of water to be tested or sold, however, these bottles indicate an amplitude and power out there in the wide world: flowing rivers, the ocean, ancient springs that are a crucial source of life for many creatures. Dion's cabinet functions as a conduit and mentally rockets one from Philadelphia to far-flung places; it is implicated in voyaging and distances. Even though this cabinet is a containing structure housing objects which are themselves containers, it opens up to wildness, to world-shaping forces, to water as both primal nutrition and supple beauty, and to the complex symbolism of water, spanning every culture and reaching way back into human history.

Collecting a water sample from the Wilmington River in Savannah.

Ralph Waldo Emerson, 1803 -1882

MICO CHLUCCO the LONG WARIO
or KING of the SIMINOLES
DO NOT OPEN
HATE
ARCHIVE

3. A Cabinet of Hate

One thing you note in William Bartram's account of his travels is how aware he was, while traveling, of the oftentimes tense relationship between Native Americans and intruding settlers or traders, and usually his sympathies are squarely with the former. The nature he explored with such enthusiasm had a rich cultural history, predating European settlement by thousands of years, and it was also a profoundly contested place, which Bartram well understood. Among the more vivid episodes in the book is one when Bartram encounters an obviously angry Native American man in a forest, a man who, it turns out, had been badly mistreated at a trading post, and who, upon escaping, had vowed "to kill the first white man he met." Bartram was unarmed and the other man had a rifle. Here's how the meeting goes:

> I never before this was afraid at the sight of an Indian, but at this time, I must own that my spirits were very much agitated: I saw at once, that, being unarmed, I was in his power; and having now but a few moments to prepare, I resigned myself entirely to the will of the Almighty, trusting to his mercies for my preservation; my mind then became tranquil, and I resolved to meet the dreaded foe with resolution and cheerful confidence. The intrepid Siminole stopped, suddenly, three or four yards before me, and silently viewed me, his countenance angry and fierce, shifting his rife from shoulder to shoulder, and looking about instantly on all sides. I advanced towards him, and with an air of confidence offered him my hand, hailing him, brother; at this he hastily jerked back his arm, with a look of malice, rage, and disdain, seeming every way discontented; when again looking at me more attentively, he instantly spurred up to me, and with dignity in his look and action, gave me his hand.

Speaking of contested places, while on his own travels Mark Dion noticed and collected virulently racist objects—really horrific stuff from a region marked by slavery, Jim Crow laws, segregation, and bigotry. He didn't have to go out of his way to find this stuff for it was readily available, still on display and available for purchase. These objects—imbued with a region's and a nation's very worst legacy—have a special role in Dion's collection; they are the only objects sealed off from view, the only objects that he's taken from public circulation and hidden, presumably forever. Twelve packages of varying sizes, wrapped in brown paper, were filled with these objects, and sent to Philadelphia, where they are displayed in their own cabinet. This cabinet is at once a marvel of understatement—just several plain, unopened packages on display with their handwritten addresses—and a harrowing encounter with racial violence and oppression; in fact, it's exactly the understatement that makes it all the more harrowing. You read Dion's rubber-stamped "Hate Archive" and the postal instructions, like "Do Not Open" and "Handle With Care," and they suddenly seem chilling, as if something truly toxic and dangerous were inside. You imagine what might be in the packages, and shudder. The fact that this toxic stuff is still commonplace makes you shudder all the more. You read the return addresses, and suddenly Dion's wandering route intersects with specific places marked and marred by racism writ large. Charleston, South Carolina, the major 18th-century slave market where slaves were bought at public auctions on the street. Pensacola, Florida, where hundreds of African Americans were the victims of racially charged murders in the early 20th century. New Orleans, with its own notorious slave market and history of lynching, especially in adjacent Jefferson Parish. Chiefland, Florida, next to Rosewood, where an African American town was attacked and burned by white mobs in 1923.

As one stares at Dion's unopened packages, they seem extremely powerful and evocative, summoning not only the South's tormented past, but enduring racial hatred, oppression and bigotry which continue to this day.

It has been a distinct pleasure to encounter a new post office and staff each few days. This is perhaps more true since I normally frequent the worst post offices in the country in New York City, where ignorance, sloth, and incompetence are continuously exhibited. Not so for the Southern post offices which are often staffed by courteous, efficient and gregarious employees. Every other day the posting of a Hate Archive box or water sample or postcard stimulates a lively conversation.

Mark Dion
Travels of William Bartram
University of West Florida
Art Dept 11000 University Parkway, Penn 32514

M. Dion
Bartram's Garden
54th St. & Lindbergh Blvd.
Philadelphia, PA
19143

DO NOT OPEN

HATE
ARCHIVE

FIRST CLASS

DO NOT OPEN

M. Dion
Bartram's Garden
54th St. & Lindbergh Blvd.
Philadelphia, PA
19143
$2.66
DO NOT OPEN
HATE
ARCHIVE

TO: M. Dion
Bartram's Garden
54th St. & Lindbergh Blvd.
Philadelphia, PA
19143
FIRST CLASS
DO NOT OPEN
FRAGILE

Mark Dion
Travels of William Bartram
M. Dion
Bartram's Garden
54th St. & Lindbergh Blvd.
Philadelphia, PA
19143
$1.81
HATE
ARCHIVE
DO NOT OPEN

Mark Dion
Travels of William Bartram
Mark Dion
Bartram's Garden
54th St & Lindbergh blvd.
Philadelphia, PA
19143
$5.20
HATE
ARCHIVE
DO NOT OPEN

Bartram's Garden
54th St. & Lindbergh Blvd
Philadelphia, PA
19143
$2.02
HATE
ARCHIVE
FIRST CLASS

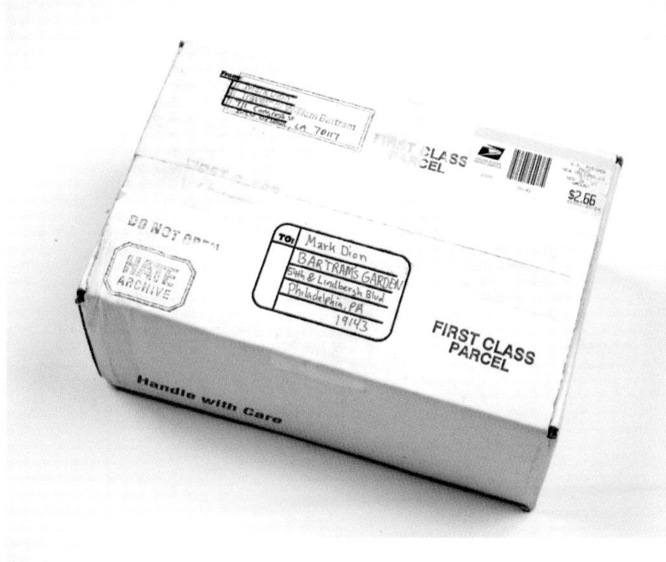
FIRST CLASS
PARCEL
$2.66
DO NOT OPEN
TO: Mark Dion
BARTRAMS GARDEN
54th & Lindbergh Blvd
Philadelphia, PA
19143
HATE
ARCHIVE
FIRST CLASS
PARCEL
Handle with Care

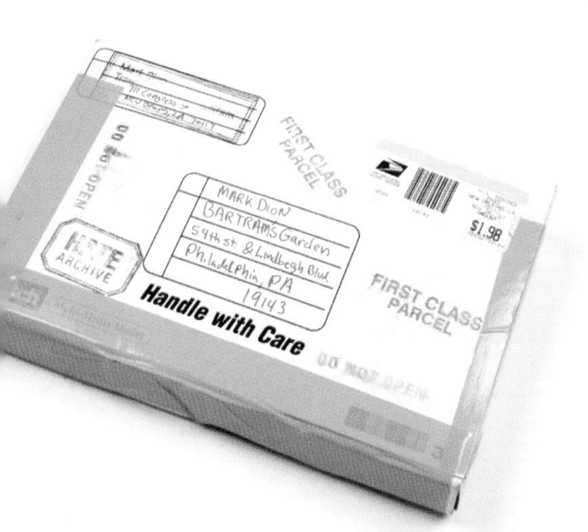
FIRST CLASS
PARCEL
DO NOT OPEN
MARK DION
BARTRAMS Garden
54th st. & Lindbergh Blvd.
Philadelphia, PA
19143
$1.98
HATE
ARCHIVE
Handle with Care
FIRST CLASS
PARCEL
DO NOT OPEN

Mark Dion
Travels of William Bartram
M. Dion
Bartram's Garden
54th St. & Lindbergh Blvd.
Philadelphia, PA
19143
DO NOT OPEN
HATE
ARCHIVE
FIRST CLASS
DO NOT OPEN

Mark Dion
Travels of William Bartram
M. Dion
Bartram's Garden
54th st & Lindbergh blvd.
Philadelphia, PA
19143
$2.83
HATE
ARCHIVE
DO NOT OPEN

4. Cabinets of Curiosities

In a speeded-up era, Mark Dion habitually slows things down; he invites patient discovery and contemplation, meaning that his work is about as far from glitz, glamour, and flashiness as you can get. In this exhibition, the more time one spends, the more drawers one opens and examines, and the more one explores the many connotations and associations provoked by the contents of those drawers, the greater the effect is, to the point where the whole exhibition becomes a voyage of discovery for the viewer. A spectacular new cabinet designed to look old features jars and vials on its top, which include a snake, eel, bugs, plants, pods, and various other preserved plantlife and animals. It takes a while to notice that these glass vessels are actually peanut butter jars, olive jars, pepper containers, and what have you from the supermarket, thus fusing botany, zoology, rampant commercialism, and garbage. As you open each of the twelve drawers, here are just a few of the things you find: plastic caps, bottle caps, pencils, tools, pincushions and old buzzers; old small spoons and ceramic fragments, old lures and rubber fish; keys, locks, safety pins and faucet handles; thimbles, toy guns, umbrellas for tropical drinks, and magic snakes; a rabbit's foot, tiny hammers, golf balls, lotions and soap from hotels, a Mickey Mouse. This entire collection once again scrambles old and new, heirlooms and forgettable debris, special things and whatever caught Dion's eye, and what's odd is how alluring this weird archive really is, visually, conceptually, and poetically. You're introduced to a dense mesh of practical actions (shaving, washing, mending, fixing) and vacation pleasures (fishing, cocktails, golf). Origins of the tourist industry abut signals of a consumerist, throwaway society, while highly mediated nature (Mickey, magic snakes, a rabbit's foot, small barnyard animals, a pink flamingo) mixes with toys and entertainment. Old tools and ceramic fragments are the heirs to homespun, pioneering days; a rebel flag shot glass reminds one of the Civil War; ice cream scoops recall vacation boardwalks; and bottle caps suggest convenience stores. You get the feeling that America, old and new, is on display, with all its convulsions and pleasantries, its mighty myths and retirement communities, its idealism and shopping.

Other cabinet drawers filled with flora and fauna are similarly eclectic. Here one finds seashells, a seahorse, a sand dollar and coral; more seashells in a densely packed grid leading from the tiniest in front to much bigger ones in the back; vials filled with seeds and pods; bark, wood, stalks, dead birds, and peanut shells. Oceans and farmland, distant forests and town trees and bushes all intermingle, in arrangements of objects which are meticulously sculptural and also strangely beautiful; indeed, a hallmark of Dion's work is how he's able to tease such a pronounced visual poetics from the most unlikely of materials. Moreover, in Dion's homemade version of a museological classification system all hierarchical distinctions between objects are suspended. Recent and remote past, valuable treasures and worthless smithereens, whole and part all flow together, and no distinction is made, for instance, between a rare specimen Dion traveled far for and a common one he casually found right outside on the front lawn. Essentially, in Dion's work, just about anything, no matter how seemingly ignoble, can have a lasting import, which implies a radical openness and has a richly democratic streak. Pointing toward this kind of attitude, Emerson wrote, in *The Poet*, "Thought makes everything fit for use. The vocabulary of an omniscient man would embrace words and images excluded from polite conversation. What would be base, or even obscene, to the obscene, becomes illustrious, spoken in a new connexion of thought." The simmering and then boiling Walt Whitman took this to heart, in poems that notice and absorb everything, no matter how spectacular or routine, and especially in his elastic, sinewy lists that are the verbal equivalent of cabinets crammed with objects. The great section 33 of *Song of Myself*, for example, is a single sentence, stretching over four pages, that moves across

Scott, Filip, Raven, Dana and I met at 8:30 to head to St. Augustine, where Bartram visited in 1765 and again in 1774, I believe. However, our first stop was at the Farmers' Market and Flea Market near Deland on Route 44. This was a real flea market and the shopping was quite good. We weaved through the aisles and I purchased numerous odds and ends for my project. Now that I have seen the Bartram cupboards I know more or less the scale of objects I must collect.

We spent the 17th and 18th in Charleston visiting the marvelous aquarium and the fantastic cemeteries which were full of botanical wonders. However the flea market just outside of town was a flop. Featuring mostly tube socks and cheap Chinese tools, this was a doomed trip. There was one exceptional table however with lots of strange Freemason ephemera and other historic oddities.

a continent to encompass log huts, lumbermen, a panther, an alligator, a black bear and buckwheat, a quail and a bat, a hot air balloon, a wrecked ship, a printing press, a shark fin, a copulating cock and hen, a Quaker woman, a moccasin print, and a baseball game, among other things. It seems to me that such an expansive, Whitmanic drive—an urge to locate and compile a wildly democratic array of materials that subverts distinctions between high and low, significant and insignificant, exalted and humble—is another key to Mark Dion's work.

"Thought makes everything fit for use. The vocabulary of an omniscient man would embrace words and images excluded from polite conversation. What would be base, or even obscene, to the obscene, becomes illustrious, spoken in a new connexion of thought." – Ralph Waldo Emerson, *The Poet*

Heineken
Beer
MOET
Corona
Extra
BUSCH
FRESH SEAL CAP
SIERRA NEVADA
USE BOTTLE OPENER
DARK
FOODS INC

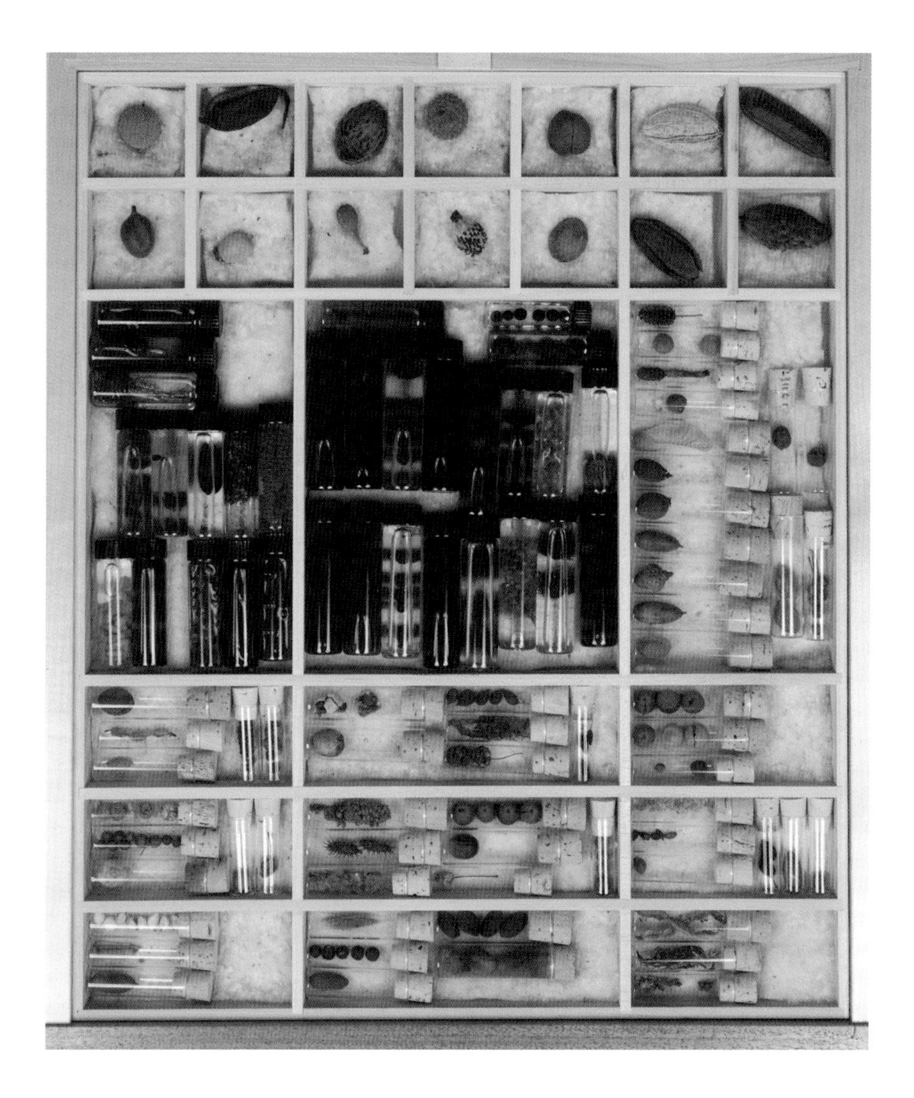

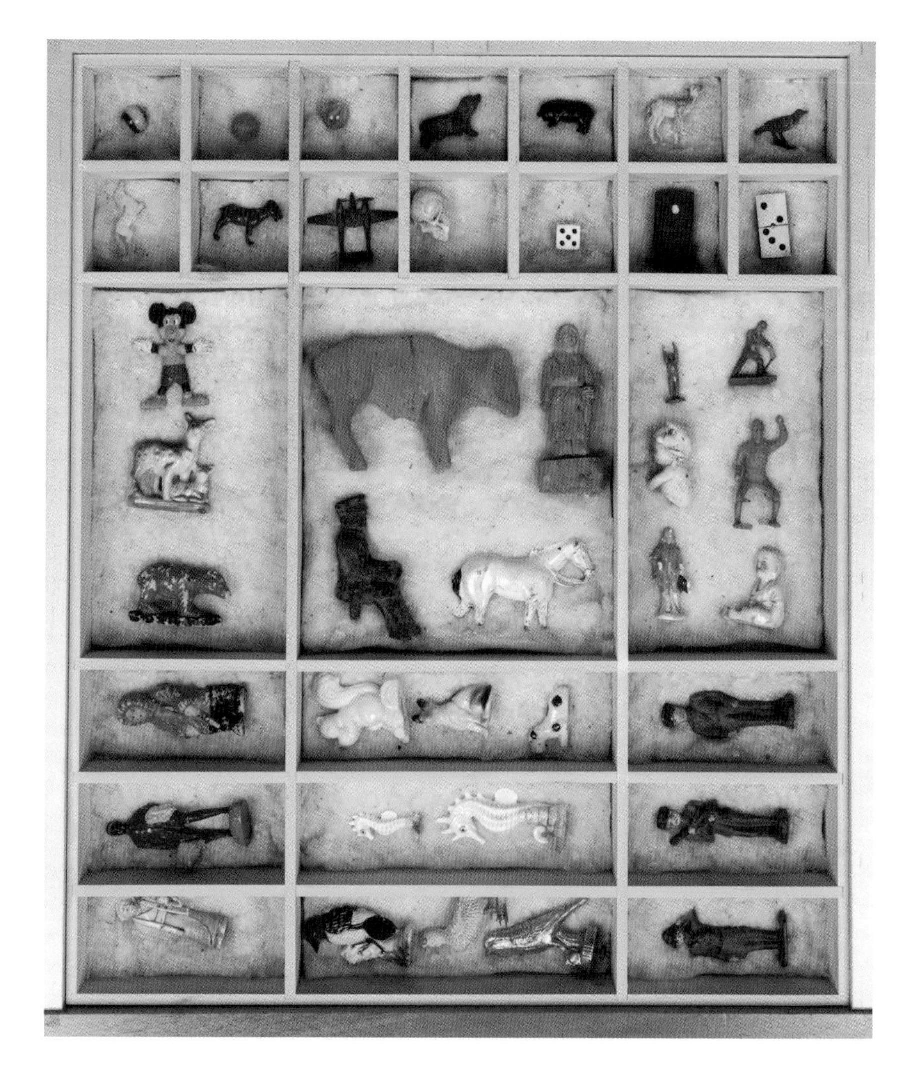

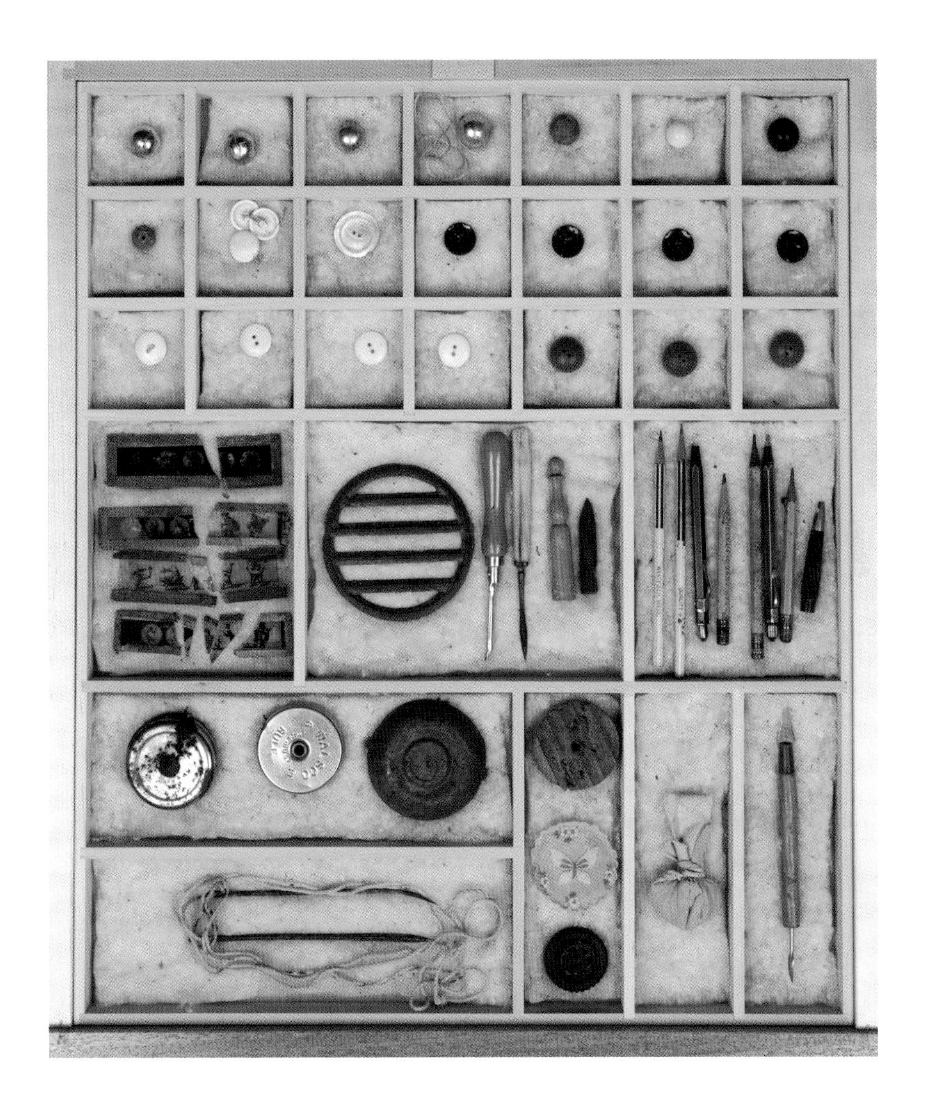

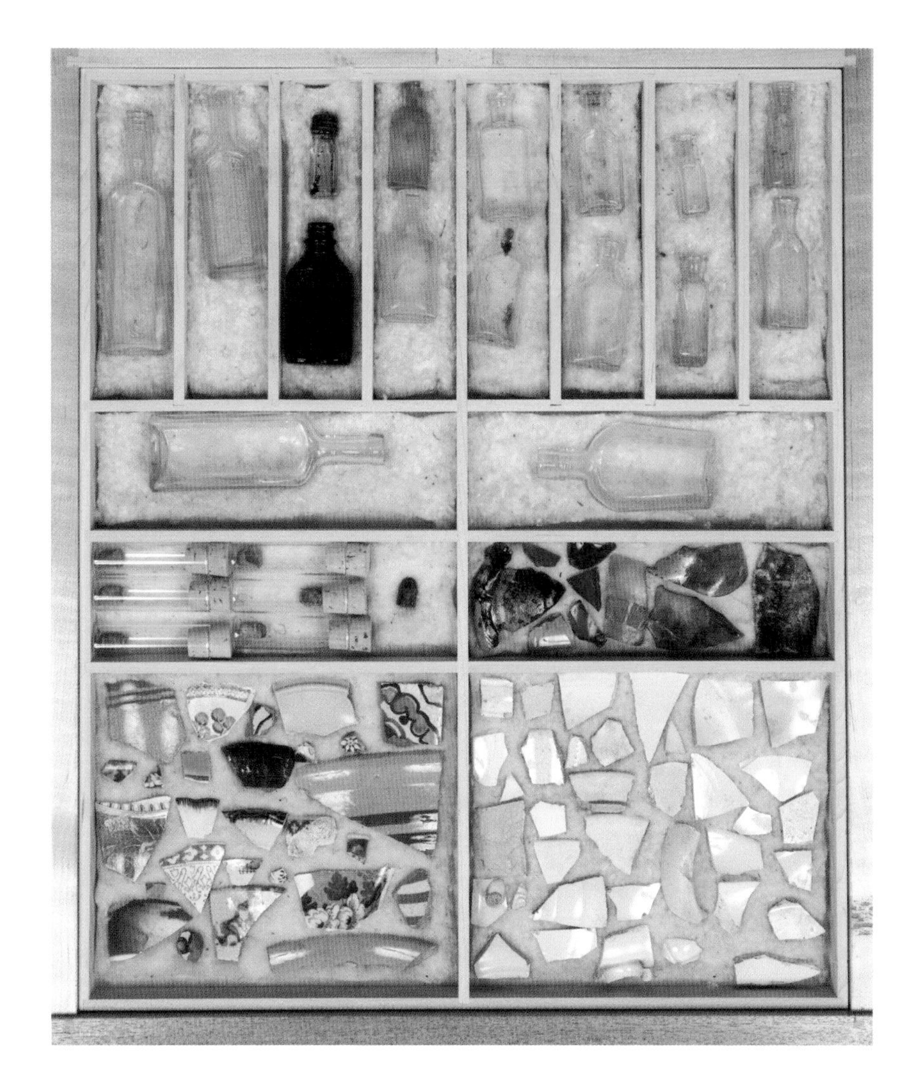

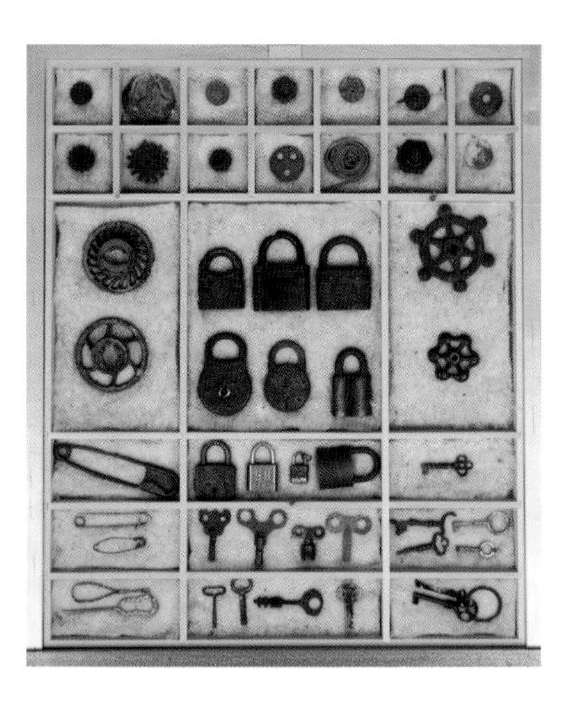

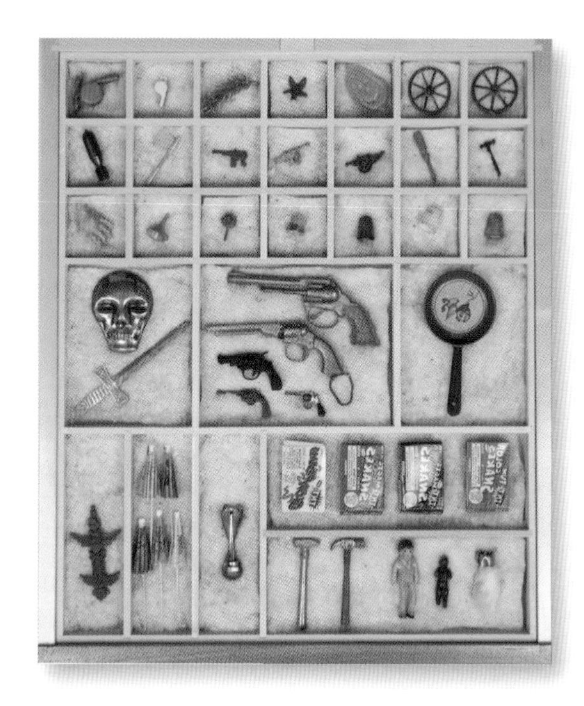

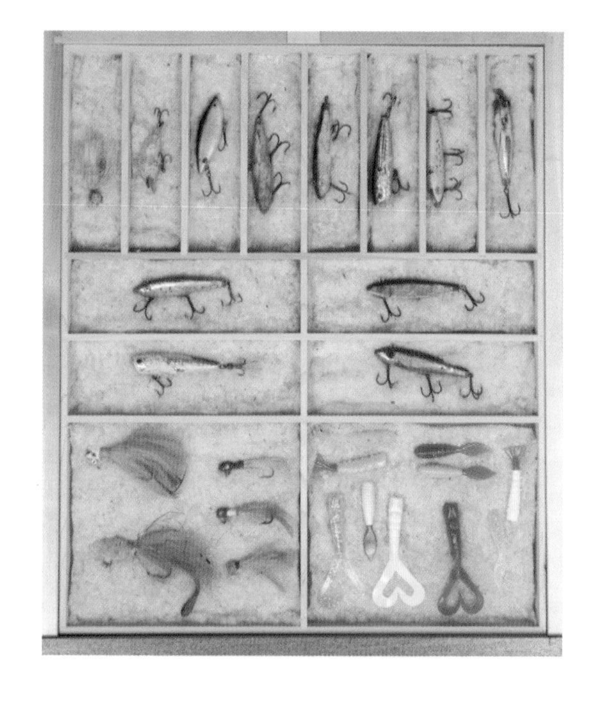

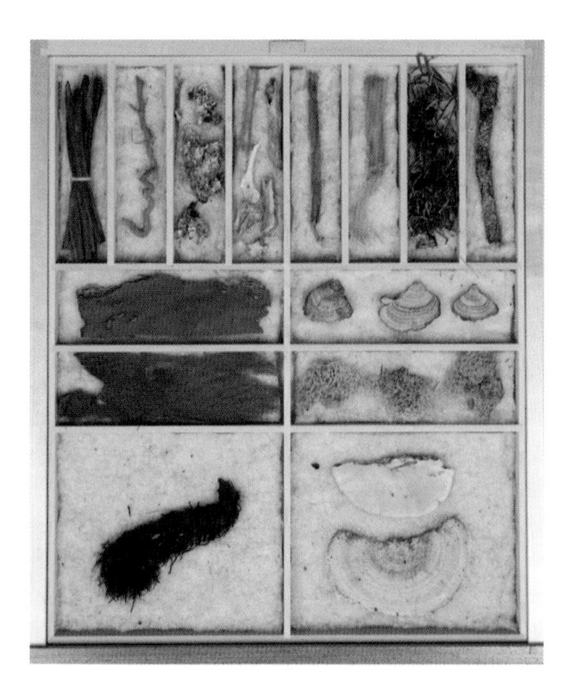

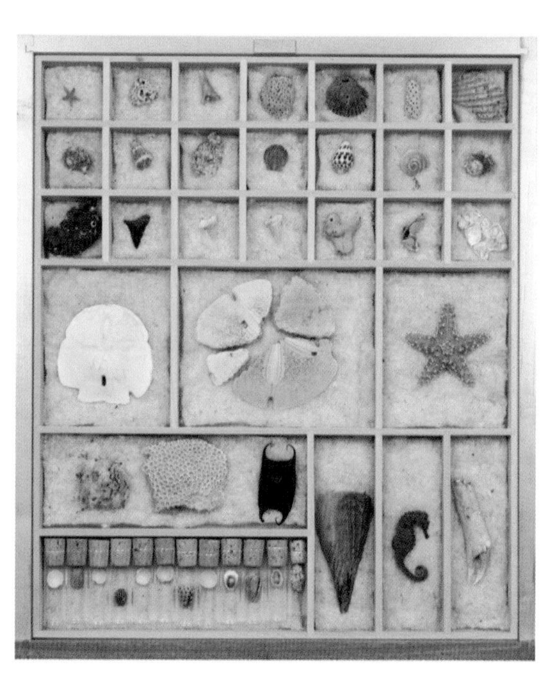

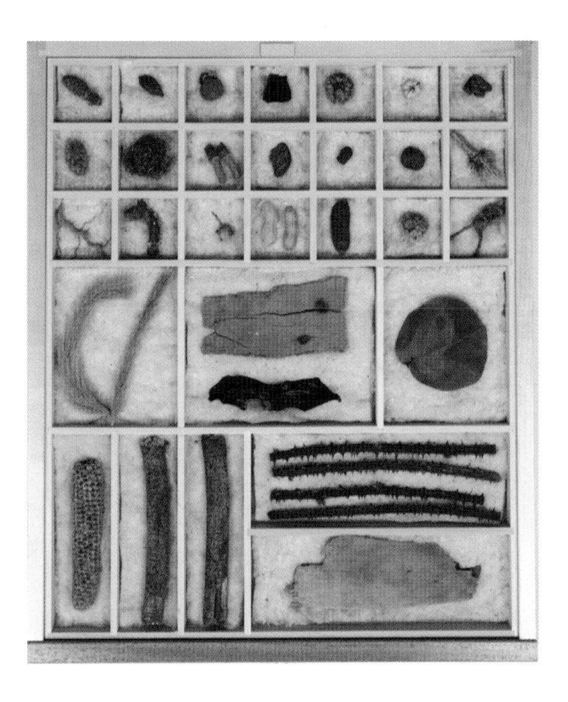

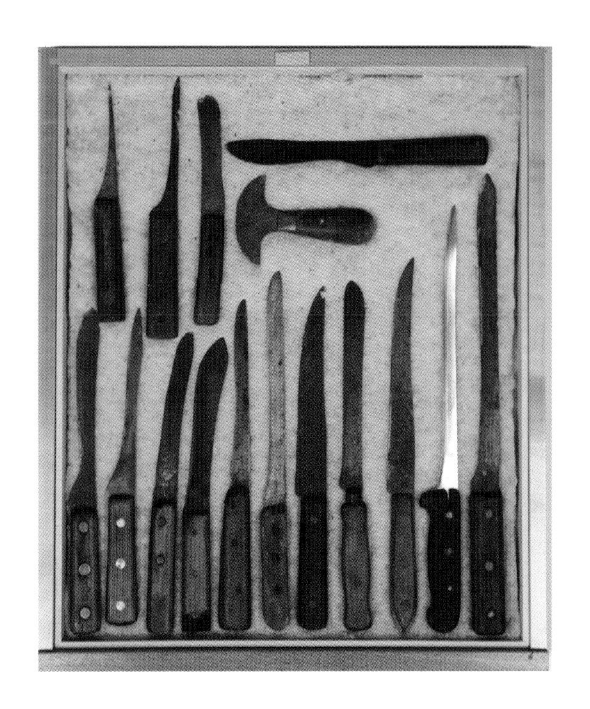

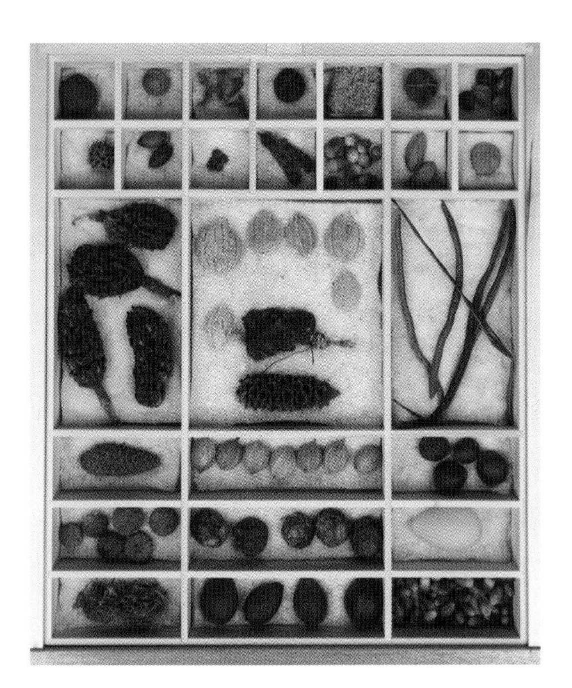

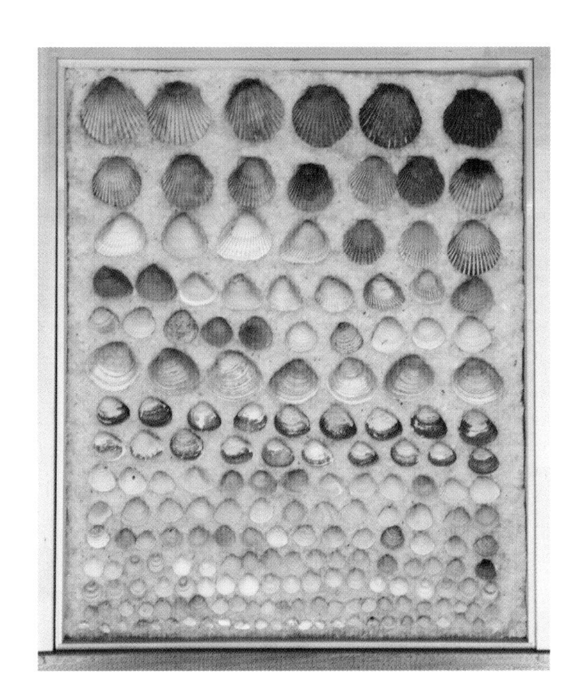

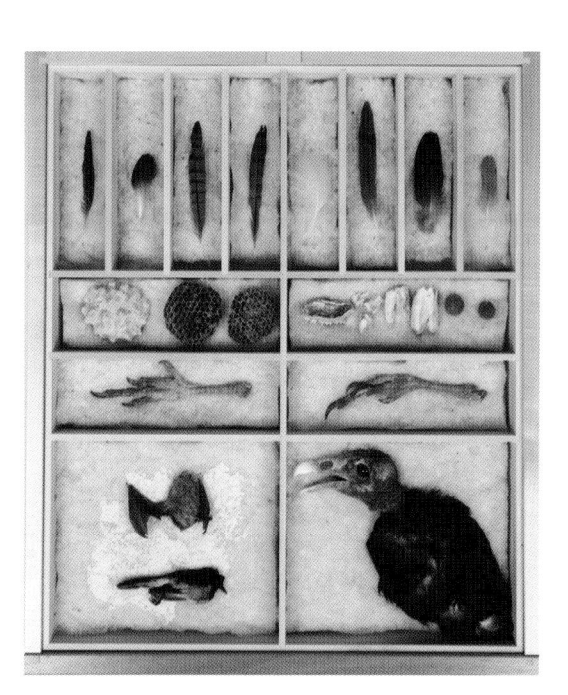

Mark Dion
Travels of William Bart

A Reddish Egret (Egretta rufescens) seen in the Mangroves of Indian River North on 10/9/08, while at the Alantic Center for the Arts.

OCT 0 8 2008

Susan Glassman
Director
Wagner Free Instite of Science
Bartram's Garden
54th st. & Lindbergh blvd
Philadelphia, PA
19143

43/9999

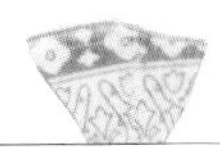

5. A Cabinet of Postcards

The Southeast, where Mark Dion traveled, and especially Florida, where he spent a great deal of time, is known for its vacation and tourist attractions. Florida, for instance, is probably the capital of postcards, to which Dion alludes, with his own cabinet of unique, hand-painted postcards. This cabinet is brand new, but built in a manner that recalls exquisite 18th-century furniture, as well as postcard displays in boardwalk souvenir shops. Two glass-and-wood doors with slots open out and feature Dion's homemade postcards, which are wonderful, coupling images with short informative texts (and you get the feeling that if he'd so chosen, Dion could have had quite a career as a painter). You see a Southern Dogface butterfly "collected and painted by Mark Dion at the Atlantic Center for the Arts, New Smyrna Beach, Florida, on the 17th day of April 2008." You see a Great Egret observed and painted in New Orleans, a Zebra butterfly, a single leaf from the plant William Bartram called China Briar, or Smilax Pseudo-China, which Dion found and painted in New Smyrna Beach. Dion's and Bartram's voyages intersect: they visited some of the same places, saw and recorded some of the same things. Dion's small, exquisite paintings also recall Bartram's excellent drawings, as well as paintings by other naturalists such as Audubon, although they contain contemporary surprises. As you look at elegant paintings of birds and butterflies, you also see a housefly, a rubber Mickey Mouse, and "A security vehicle from the Flea and Farmer's Market outside Deland, Florida at the junction of Route 444 and Interstate highway 4. Seen and represented by M. Dion on April 23 2008." It's a hybrid place that Dion explored, and he's as interested in houseflies and motorized carts as he is in lovely butterflies.

These postcards were sent to Bartram's Garden addressed to specific people, just like normal postcards, and one imagines they were anticipated with enthusiasm and received with great delight. As he traveled through the Southeast, loosely following William Bartram's route, Dion had adventures with friends, encounters with many strangers, conducted Web conferences with high school students, was a Master Artist at the Atlantic Center for the Arts in Florida (where I visited him), fielded email inquiries, and communicated constantly with his colleagues up north at Bartram's Garden. All of this interpersonal interaction is as much a part of the whole project as anything else, as anything Dion saw and collected, and made this trip (which is typical for many of Dion's projects) not only artistically but humanly cathartic. What could have been a generally solitary excursion was actually anything but. Instead, Dion was present at all points, as an organizer, source of inspiration, canny tactician, theoretician, Barnumesque promoter, ringmaster, comedian, infectious enthusiast, and true believer. Even though the project wasn't about collaboration per se, collaboration was central to it, and it doubled as a roving meeting ground and think tank. This project moved far beyond the art world into, well, the world, and I'm quite certain that many of the participants will remember and be inspired by their experience for a good long while. Without fanfare, and arising more from a spirit of adventure and generosity than from any ideology, Dion's project inspired a temporary community to come together in common purpose, with commitment, as well as with risk, joy, driving inquiry, and occasional hilarity. Mark Dion is a brilliant artist, yet one reason why his projects are so compelling has everything to do with their palpable humanity and their loopy utopianism. They consistently cross over from an arts context into normal, daily life, and they do so with profundity, gleeful ridiculousness, and with an aptitude for renovation and renewal.

Dion tucks into his journal for an in situ account of the visit to St. Augustine.

Ring-billed Gull, seen on the Lake Pontchartrain Causeway by Mark Dion. Louisiana NOV 25 2007

William Bartram sailed along the northern shore of Lake Pontchartrain in mid October 1775.

Stephanie Phillips
BARTRAM'S GARDEN
54th st & Lindbergh Blvd.
Philadelphia, PA
19143

Cotman Water Colour Paper
Papier aquarelle. 140lbs (300gsm)

A Ceramic Fragment Collected from the graveyard of the Saint Philips Episcopal church on Church st, Charleston, South Carolina by M. Dion. NOV 17 2007

William Bartram arrived in the city of Charleston on March 31, 1773.

Julie Courtney
Bartram's Garden
54th Street & Lindbergh Blvd.
Philadelphia, PA 19143

Cotman Water Colour Paper
Papier aquarelle. 140lbs (300gsm)

Mark Dion
Travels of William Bartram

GAINESVILLE GNV
FL 326
PM 3 L

An insect damaged leaf collected in Manatee Springs State Park, Florida, by M. Dion.

DEC 04 2007

William Bartram traveled to Manatee Springs in June 1774 on a day excursion down the Sowanne River from Talahasochte. He saw skeletons of Manatee killed and feasted on by the Indians.

Kendra Gaeta
Bartram's Garden
54th st and Lindbergh Blvd.
Philadelphia, PA
19143

Mark Dion
Travels of William Bartram

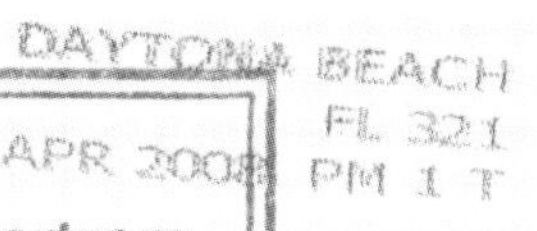

A Marsh Periwinkle (Littorina irrorata (Say)) collected on brackish grasses in Strickland Bay, Port Orange Florida on the 18th of April 2008.
M. Dion

Louise Turan
Bartram's Garden
54th st & Lindbergh Blvd.
Philadelphia, PA
19143

+612a C063

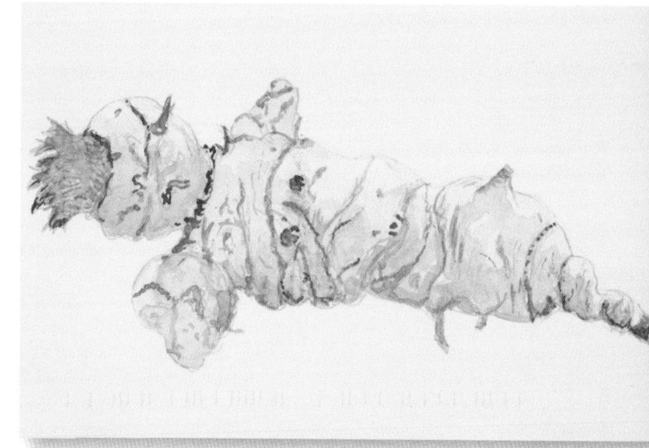

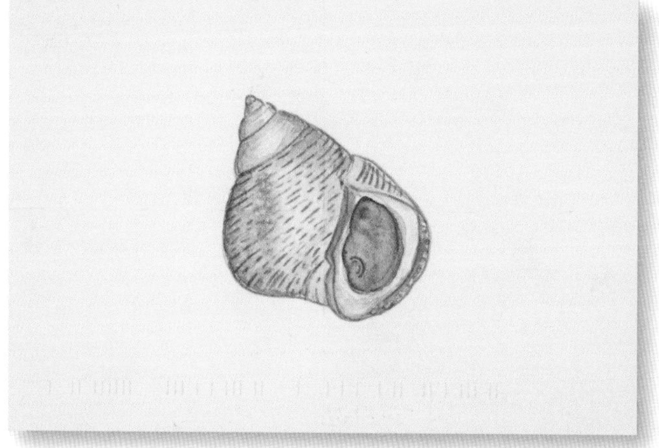

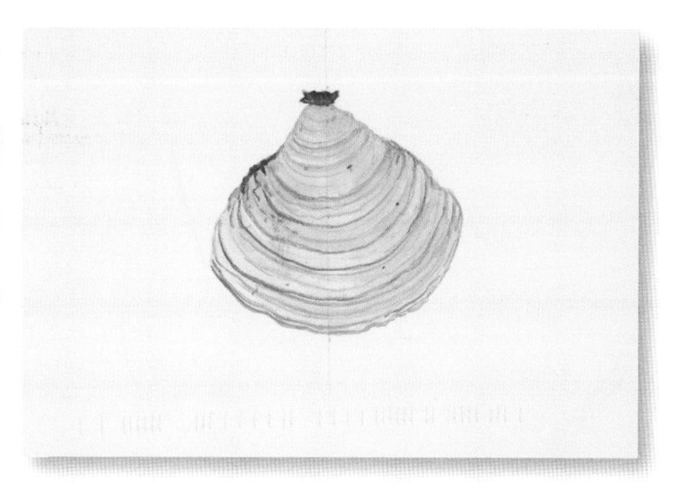

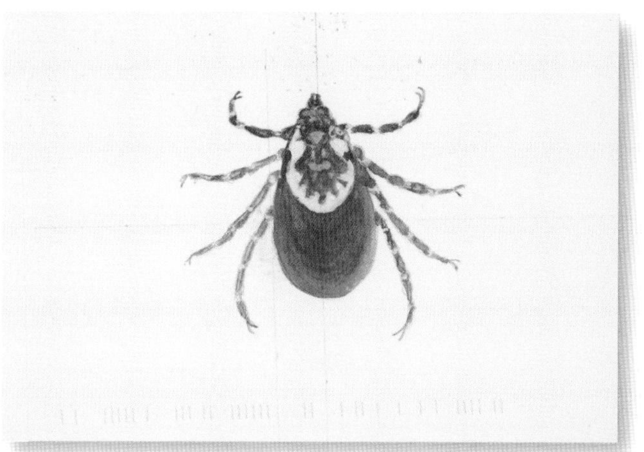

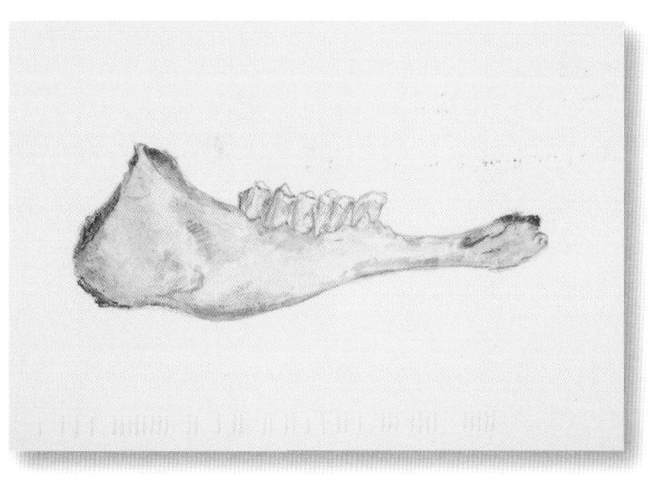

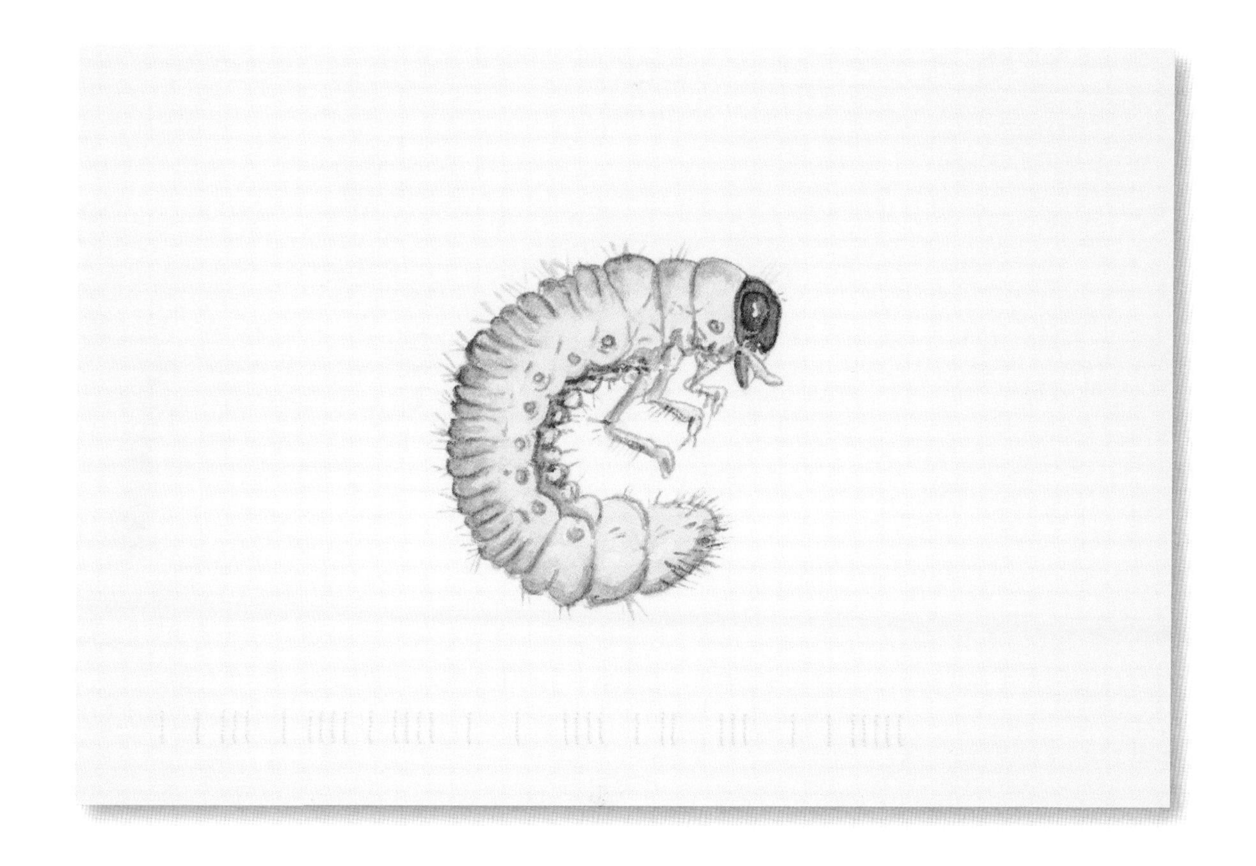

The original *Travels...* by William Bartram

Travels Through North & South Carolina, Georgia, East & West Florida,... the Cherokee Country, the Extensive Territories of the Muscogulges, or Creek Confederacy, and the Country of the Chactaws; Containing An Account of the Soil and Natural Productions of Those Regions, Together with Observations on the Manners of the Indians. Embellished with Copper-Plates. Philadelphia: James & Johnson, 1791.
The entire 1791 edition is available online at the Documenting the American South project if UNC-CH: http://docsouth.unc.edu/nc/bartram/menu.html

Works by John Bartram

The Correspondence of John Bartram 1734–1777. Edited by Edmund Berkeley and Dorothy Smith Berkeley. Gainesville: University Press of Florida, 1992.

"Diary of a Journey through the Carolinas, Georgia, and Florida, from July 1, 1765 to April 10, 1766." Edited by Francis Harper. *Transactions of the American Philosophical Society*, n.s. vol. 33, part 1 (December 1942), pp. 1–122.

"A Journal Kept by John Bartram of Philadelphia: Botanist to His Majesty for the Floridas; upon a journey from St. Augustine up the River St. Johns." In William Stork, *An Account of East-Florida, with a journal kept by John Bartram of Philadelphia...* 2nd edition, London, 1767; 3rd ed., 1769; 4th ed., 1774.

Memorials of John Bartram and Humphry Marshall. Edited with introduction by William Darlington. Philadelphia: Lindsay & Blakiston, 1849.
Reprint edition with introduction and chronology by Joseph Ewan. New York: Hafner Publishing Company, 1967.

Observations on the Inhabitants, Climate, Soil, Rivers, Productions, Animals, and Other Matters Worthy of Notice Made By John Bartram, in his Travels from Pensilvania to Onondaga, Oswego, and the Lake Ontario in Canada... London, 1751.

Works by William Bartram

A Catalogue of Trees, Shrubs, and Herbaceous Plants, Indigenous to the United States of America; Cultivated and Disposed of By John Bartram & Son, At their Botanical Garden, Kingsess, near Philadelphia: To Which is Added A Catalogue of Foreign Plants, Collected From Various Parts of the Globe. Philadelphia: Bartram and Reynolds, 1807.

"Travels in Georgia and Florida, 1773–74. A Report to Dr. John Fothergill." Edited by Francis Harper. *Trans. of the American Philosophical Society*, n. s. vol. 33, part 2 (Nov. 1943), pp. 121–242.

Travels and Other Writings. Edited by Thomas P. Slaughter. New York: The Library of America, 1996.

Travels of William Bartram. Edited by Mark van Doren. 1928; reprint New York: Dover, 1955.

The Travels of William Bartram: Naturalist's Edition. Edited by Frances Harper. New Haven, CT: Yale University Press, 1958.

Travels Through North & South Carolina, Georgia, East & West Florida,... the Cherokee Country, the Extensive Territories of the Muscogulges, or Creek Confederacy, and the Country of the Chactaws; Containing An Account of the Soil and Natural Productions of Those Regions, Together with Observations on the Manners of the Indians. Embellished with Copper-Plates. Philadelphia: James & Johnson, 1791.
The entire 1791 edition is available online at UNC-CH Documenting the American South project: http://docsouth.unc.edu/nc/bartram/menu.html.

William Bartram Botanical and Zoological Drawings, 1756–1788. Edited and annotated by Joseph Ewan. Philadelphia: American Philosophical Society, 1968.

William Bartram on the Southeastern Indians. Edited and annotated by Gregory A. Waselkov and Kathryn E. Holland Braund. Lincoln: University of Nebraska Press, 1995.

Works about the Bartrams, Mark Dion, Museums, Collections, and Display

Bartram Trail Conference, Links to numerous Bartram-related Websites, http://www.bartramtrail.org/pages/links.html

Bennett, Tony. *The Birth of the Museum: History, Theory, Politics.* London and New York: Routledge, 1995.

Bryllion, Fagin N. *William Bartram, Interpreter of the American Landscape.* Baltimore: The Johns Hopkins Press 1933.

Buchhart, Dieter, and Verena Gamper et al. *Mark Dion: Concerning Hunting.* Ostfildern, Germany: Hatje Cantz, 2008.

Buck, Rebecca, and Jean Allman Gilmore. *Collection Conundrums: Solving Collections Management Mysteries.* Washington, D.C.: American Association of Museums, 2007.

Coles, Alex, and Mark Dion, eds. *Mark Dion: Archaeology.* London: Black Dog, 1999.

Corrin, Lisa, Miwon Kwon, and Norman Bryson. *Mark Dion.* London: Phaidon, 1997.

Corrin, Lisa, ed. *Mining the Museum: An Installation by Fred Wilson.* Baltimore: the Contemporary; New York: The New Press, 1994.

Davis, Jack E.,. and Raymond Arsenalt, eds. *Paradise Lost? The Environmental History of Florida.* Gainesville: University of Florida, 2005.

Dion, Mark. *Bureau of the Center for the Study of Surrealism and Its Legacy.* Manchester: AHRB Research Centre for Studies of Surrealism and its Legacies; London: Bookworks, 2005.

Dion, Mark. Die geöffnete Raritäten- und Naturalien-Kammer. Cologne: Salon, 2002.

———. *Ursus Maritimus (Polar Bear).* Cologne: Walther König, 2003.

Dion, Mark, and Alexis Rockman, eds. *Concrete Jungle.* New York: Juno Books, 1996.

Duncan, Carol. *Civilizing Rituals: Inside Public Art Museums*. London and New York: Routledge, 1995.

Elsner, John, and Roger Cardinal, eds. *The Cultures of Collecting* Cambridge, Mass.: Harvard University Press, 1994.

Foucault, Michel. *The Order of Things: An Archaeology of the Human Sciences*. New York: Vintage, 1994.

Grässlin, Karola, ed. *Mark Dion: Die Wunderkammer*. Munich: K-raum Daxer, 1993.

Heidemann, Christine, et al. *Mark Dion: Encyclomania*. Nuremberg, Germany: Verlag für Moderne Kunst. 2003.

Impey, Oliver, and Aurthur MacGregor, eds. *The Origins of Museums: The Cabinet of Curiosities in Sixteenth- and Seventeenth-Century Europe*. London: Clarendon Press, 1998.

Hammonds, Kit, ed. *Mark Dion: Microsomographia*. London: South London Gallery, 2005.

Herbst, Josephine. *New Green World: John Bartram and the Early Naturalists*. New York: Hasting House, 1954.

Hooper-Greenhill, Eilean. *Museums and the Shaping of Knowledge*. London: Routledge, 1992.

Karp, Ivan, and Steve D. Lavine, eds. *Exhibiting Cultures: The Poetics and Politics of Museum Display*. Washington, DC: Smithsonian Institution Press, 1991.

Kastner, Joseph. *A Species of Eternity*, New York: Knopf, 1977.

Klein, Richard, ed. *Mark Dion: Drawings, Journals, Photographs, Souvenirs, and Trophies 1990–2003*. Ridgefield, Connecticut.: The Aldrich Museum of Contemporary Art, 2003.

Magee, Judith. *The Art and Science of William Bartram.* State College: Pennsylvania State University Press; London: The Natural History Museum, 2007.

McShine, Kynaston. *The Museum as Muse: Artists Reflect*. New York: Museum of Modern Art, 1999.

Markonish, Denise, and Gregory Volk. *Mark Dion: New England Digs.* Brockton, Massachusetts: Fuller Art Museum, 2001.

Olalquiagu, Celeste. *The Artificial Kingdom: A Treasury of the Kitsch Experience*. New York: Pantheon, 1998.

Parsy, Paul-Hervé. *Mark Dion: Dungeon of the Sleeping Bear, the Phantom Forest, the Birds of Guam and Other Fables of Ecological Mischief*. Oiron, France: Chateau d'Oiron, 2005.

Peabody, William. *Life of Alexander Wilson* Boston, 1848.

Pearce, Susan M. *Museums, Objects, and Collections: A Cultural Study*. Washington, D.C.: Smithsonian Institution Press, 1993.

Pearce, Susan M., ed. *Interpreting Objects and Collections.* London and New York: Routledge, 1994.

Pomian, Krzysztof. *Collectors and Curiosities: Paris and Venice 1500–1800*. Trans. Elizabeth Wiles-Portier. Cambridge: Polity Press, 1991.

Prince, Sue Ann, ed. *Stuffing Birds, Pressing Plants, Shaping Knowledge: Natural History in North America, 1730–1860*. Philadelphia: American Philosophical Society, 2003.

Pugnet, Natacha. *Mark Dion: The Natural History of the Museum*. Paris: Archibooks, 2007.

Putnam, James. *Art and Artifact: The Museum as Medium.* London: Thames and Hudson, 2001.

Quimby, Ian M.G., ed. *Material Culture and the Study of American Life.* New York: W.W. Norton and Winterthur Museum, 1978.

Sanders, Brad. *Guide to William Bartram's Travels: Following the Trail of America's First Great Naturalist*. Athens, Georgia: Fevertree Press, 2002.

Sheehy, Colleen J., ed. *Cabinet of Curiosities: Mark Dion and the University as Installation*. Minneapolis: University of Minnesota Press and the Weissman Art Museum, 2006.

Slaughter, Thomas P. *The Natures of John and WIlliam Bartram*. New York: Vintage Books, 1998.

"Some Account of the Late Mr. John Bartram, of Pennsylvania." *Philadelphia Medical and Physical Journal*, vol. 1, part 1 (1804), p. 115–24.

Spornick, Charles D., Ian R. Cattier, Robert J. Greene. *An Outdoor Guide to Bartram's Travels*. Athens: University of Georgia Press, 2003.

Wallach, Alan. *Exhibiting Contradiction: Essays on the Art Museum in the United States.* Amherst: University of Massachusetts Press, 1998.

Weschler, Lawrence. *Mr. Wilson's Cabinet of Wonder: Pronged Ants, Horned Humans, Mice on Toast, and Other Marvels of Jurassic Technology.* New York: Pantheon, 1995.

"William Bartram (1791)," *Southern Nature: Scientific Views of the Colonial American South.* American Philosophical Society, 20001, http://www.amphilsoc.org/library/exhibits/nature/bartram.htm

Credits

Travel diary and location photography: copyright © Dana Sherwood
(unless otherwise indicated).

Exhibition and artifact/object photography: copyright © Aaron Igler
(unless otherwise indicated).

Exhibition cabinetry: Ian and Matt Pappajohn and Kristine Kennedy of Pappajohn Woodworking.

Page 2: Oakleaf hydrangea, *Hydrangea quercifolia*, engraving based on William Bartram drawing, from *Travels*, Philadelphia: 1791. The John Bartram Association, Bartram's Garden, Philadelphia.

Page 5: Houses and lawn alligator: Dana Sherwood; Pingpong players: Scott Hocking;
Pottery shards: Aaron Igler.

Page 11: Leaves: Dana Sherwood.

Page 12: Detail, Hate Archive cabinet: Jeffrey Jenkins.

Page 13: Artist's tools: Dana Sherwood.

Page 14: Stamp illustration: Jeffrey Jenkins.

Page 15: Seeds: Dana Sherwood.

Page 18: *A Draught of John Bartram's House and Garden as it appears from the River.* The only contemporary plan of John Bartram's Garden, this drawing was probably made by William Bartram in mid- to late 1758, and was sent to Peter Collinson in London, in a letter dated January 28, 1759. William captured his father's garden in a state of expansion, with a new flower garden and major alterations to the house just underway. The figure standing among the woody plant collection in the alleys of the lower garden is probably John himself. Collection of the Earl of Derby, Knowsley Hall. (The Earl of Derby's Library at Knowsley Hall, near Liverpool, owns a large folio volume of drawings, once owned by Peter Collinson, John Bartram's friend and chief correspondent in London. The volume contains 30 or so illustrations by William Bartram, most produced when he was young. It also includes this much reproduced plan of Bartram's Garden.) License granted courtesy of the Rt Hon The Earl of Derby 2008.

Page 19: Bartram house and garden: Jeffrey Jenkins; Title page: *A Catalogue of Trees, Shrubs, and Herbaceous Plants, Indigenous to the United States of America...* Philadelphia: 1807. Collection of the Library Company of Philadelphia.

Page 20: *William Bartram*, portrait in oil by Charles Willson Peale, 1808. Original at the Portrait Gallery, Independence National Historical Park, Philadelphia, National Park Service.

Page 21: **1**: Nicholas Scull and George Heap, *A Map of Philadlephia and Parts Adjacent...*, engraved by Lawrence Herbert, Philadelphia: 1752. There are many versions and later reprints of this map. The most common is from London ca. 1777. The original is a better source as it spells "Bartram" correctly (circled). Library of Congress, Geography and Map Division. **2**: Philadelphia *Aurora* newspaper ad for Bartram's Garden. The John Bartram Association, Bartram's Garden, Philadelphia. **3**: Postcard of window inscription. The John Bartram Association, Bartram's Garden, Philadelphia. **4**: Title page, Linnaeus, *Genera Plantarum*, 2nd edition, Leiden: 1742. John Bartram recieved this book from Johan Frederik Gronovius, a botanist in Leyden, The Netherlands, in 1743. Signed flyleaf, front and rear: Signed "John Bartram His Book given him by Dr. Gronovius chief Profesor at Leiden 1743" and "W. Bartram His Book June the 10 1755". Dr. John B. Bartram Special Collections Library, Bartram's Garden. The John Bartram Association, Bartram's Garden, Philadelphia.

Page 22: Portrait of Peter Collinson, The John Bartram Association, Bartram's Garden, Philadelphia.
Black-throated Green Warbler (*Dendroica virens*), on Red Oak (*Quercus rubra*)". License granted courtesy of the Rt Hon The Earl of Derby 2008.

Page 23: *Old Florida Cypress Bartram's Garden*, ca. 1875. Photograph by John Moran. The John Bartram Association, Bartram's Garden, Philadelphia.
Details: east facade of the Bartram house: Jeffrey Jenkins.

Page 24: Franklin tree, *Franklinia alatamaha*, William Bartram drawing engraved by James Trenchard, Philadelphia, ca. 1786. Courtesy of the American Philosophical Society.
Seed packets: Jeffrey Jenkins.

Page 25: Seeds and letter photograph: Jeffrey Jenkins.
Catalogue D'Arbres D'Arbustes et de Plantes... Paris: 1783. Original, American Philosophical Society.

Page 26 (and page 112): *Catalogue of American Trees, Shrubs and Herbacious Plants...* Philadelphia: 1783. Original, Bartram Papers, Historical Society of Pennsylvania.
Footbridge across the Philadelphia, Wilmington, and Baltimore Rail Road. The John Bartram Association, Bartram's Garden, Philadelphia.

Page 27: East facade of Bartram House, ca. 1882. Photograph: William H. Rau. The John Bartram Association, Bartram's Garden, Philadelphia.
Red Owl, Warbling Flycatcher, Purple Finch, Brown Lark, Alexander Wilson, American Ornithology, Philadelphia: 1812, vol. 5, plate 42. The John Bartram Association, Bartram's Garden, Philadelphia.

Page 28: Old cider mill on the bank of the Schuylkill, 1892. Unknown photographer. The John Bartram Association, Bartram's Garden, Philadelphia.
Contemporary view of cider mill, 2008: Jeffrey Jenkins.

Page 29: Bartram house and garden: Jeffrey Jenkins
A Draught of John Bartram's House and Garden as it appears from the River. (see page 18 credit)

Page 30-31: 1-5: The John Bartram Association, Bartram's Garden, Philadelphia.
Page 34: Title page of William Bartram's *Travels* , 1792 London. The John Bartram Association, Bartram's Garden, Philadelphia.

William Bartram — A Portfolio
Page 38: *"Motacilla. A Sollaterry Bird,"* Louisiana Waterthrush, *Seiurus motacilla, and "Yellow Spiked Lycimacha,"* swamp candles, *Lysimachia terrestris*. License granted courtesy of the Rt Hon The Earl of Derby 2008.

Page 39: *A View of the underside of the great Mud Tortoise from Pennsylvania.* Common snapping turtle *Chelydra serpentina*, watercolor & ink on paper, 1759. License granted courtesy of the Rt Hon The Earl of Derby 2008.

Page 40: *Purple Flower'd Ixia* from Florida, the celestial lily, *Calydorea coelestina*, watercolor & ink on paper, 1767. License granted courtesy of the Rt Hon The Earl of Derby 2008.

Page 41: *The great Silver-Leafed River Maple*, silver maple, *Acer saccharinum*, from a set of all the Pennsylvania maples by William Bartram sent to Peter Collinson, watercolor & ink on paper, 1755. License granted courtesy of the Rt Hon The Earl of Derby 2008.

Page 42: Two Florida plants, gopher apple, *Licania michauxii*, and hairy laurel, *Kalmia hirsuta*, drawn by William Bartram and engraved by James Trenchard, Philadelphia, 1786. Courtesy of the American Philosophical Society.

Page 43: Large whorled pogonia, *Isotria verticillata*, and rosebud orchid, *Cleistes divaricata*, ink drawing and description of two North American orchids from 1796. Courtesy of the American Philosophical Society.

Page 45: *The Great Alachua-Savana; in East Florida*, William Bartram map and nature tableau of this unique environment, created during or after his visit in 1774. Courtesy of the American Philosophical Society.

Page 46: Indian shot, *Canna indica*, William Bartram drawing from 1784. Courtesy of the American Philosophical Society.

Page 47: Fevertree, *Pinckneya bracteata*, William Bartram drawing, engraved by James Trenchard, Philadelphia, ca. 1786. Courtesy of the American Philosophical Society.

Page 50: Pingpong players: Scott Hocking.

Page 50-65: All *Mark Dion Journal Excerpts'* photographs: Dana Sherwood.

Page 66: *Alligator of St Johns*, from the botanical and zoological drawings (1756-1788) by William Bartram. The Natural History Museum, London, UK.

Page 67: Interior, alligator cabinet: Jeffrey Jenkins.

Page 72: *Franklinia alatamaha*, Franklin tree, specimen by William Bartram, 1773-76. The Natural History Museum, London, UK.

Page 73: Detail, Mark Dion's herbarium cabinet: Jeffrey Jenkins.

Page 76: Title page from John and William Bartram's copy of Linnaeus's 1742 *Genera Plantarum*. The John Bartram Association, Bartram's Garden, Philadelphia.
Detail from Dion's exhibition *Systema Metropolis* at London's Natural History Museum, 2007. The Natural History Museum, London, UK.

Page 77: Mark's herbarium cabinet: Jeffrey Jenkins.

Page 78: Water sample cabinet: Jeffrey Jenkins.

Page 79: Ralph Waldo Emerson: Dover Publications.

Page 80: Title page of William Bartram's *Travels* 1791 edition, Philadelphia. The John Bartram Association, Bartram's Garden, Philadelphia.

Page 81: Detail, Hate Archive cabinet: Jeffrey Jenkins.

Page 111: Jeffrey Jenkins.

Mark Dion, Dana Sherwood and Julie Courtney at the Bartram house, June 20, 2008.

Seed in JOHN BARTRAM's Garden, near Philadelphia. The Seed and growing Plants of which are disposed of on the most reasonable

Latin name	Common name	
LIRIODENDRON	Tulip Tree	a
Liquid Ambar styracifolia	Sweet Gum	
---Aspleni Folia	Sweet Fern	
Pinus Strobilus	White Pine	b
---Palustris	Swamp Pine	a
---Sylvestris	Pinaster or Mountain Pine	c
---Toda	Frankinsense Pine	b
---Phoenix	Long leaved Pine	
---Pumila	Dwarf Pine	
---Echinatus	Pennsylvania Pine	
---Larix	Larch Tree, White and Red	a
---Abies Canadensis	Balm of Gilead Fir	
---Abies Virginiana	Hemlock Spruce	c
---Abies Pini Foliis Brevibus	Newfoundland Spruce, black red and Dwarf	a
Cupressus Disticha	Bald Cyprus	
---Thyoides	White Cedar	
Juniperus Virginiana	Red Cedar	o
Laurus Sasafras	Sasafras Tree	o
---Nobilis	Red Bay	
---Benzoin	Benjamin or Spice Wood	a
---Geniculata	Carolina Spice Wood	
Magnolia Glauca	Rose Laurel	d
---Grandiflora	Florida Laurel Tree	o
--Acuminata	Cucumber Tree	a
---Umbrella	Umbrella Tree	
Prunus Americana	Great Yellow Sweet Plumb	o
---Missisippe	Crimson Plumb	
---Chicasaw	Chicasaw Plumb	b
---Maritima	Beach or Sea-side Plumb	a
---Declinatus	Dwarf Plumb	
---Padus Sylvatica	Bird or Cluster Cherry	o
---Serratifolia	Evergreen Bay of Carolina	a
---Racemosa	D[illegible] Bird Cherry	
Malus Coronaria	[illegible]b Apple	
[illegible]bus Americana	Quickb[illegible]am	
[illegible]pilus Nivea	Sweet Service	
---Pumila	Dwarf [illegible]weet Service	
---Azarol	Grea[illegible] [illegible]awthorn	a
---Spinoza	Cockspur Hawthorn	
---Humilis	Dwarf Hawthorn	b
---Apiifolia	Carolina Hawthorn	
Crategus Prunifolia	Swamp Service, 2 varieties with red and black Fruit	d
---Canadensis	————	
------	Dwarf Swamp Service 2 varieties red and black Fruit	a
Thuya Odorifera	Arbor Vitae	
Cornus Florida	Dogwood	
---Sylvestris		e
---Sanguinea	Red Willow	a
---Perlata	White berried Swamp Dogwood	
---Venosa	Mountain Dwarf Cornus	e
Viburnum Prunifolia	Great Black Haw	a
---Spinosum	Small black Haw	
---Tini folia	Water Elder	d
---Triloba	Mountain Viburnum	c
---Lanceolata	Tough Viburnum	a
---Dentatum	Arrow wood	
---Alnifolium		c
Sambucus	Elder	a

Latin name	Common name	
Vaccinium Arboreum	Tree Whortleberry	b
——Pendulum	Indian Goosberry	
——Evonimifolium	Swamp Whortleberry	d
——Pusillum	Whortleberry with a large black Fruit	b
——Racemosum	————	
——Nigrum	Black Whortleberry	
——Pavifolium	Billberry	
——Palustre	Oycoccus or Cranberry	d
Rhododendron maximum	Mountain Laurel	c
Kalmia latifolia	Common Laurel	
——Glauca	Dwarf Laurel	d
——Angustifolia	Thyme leav'd Kalmia	
——Ciliata	Dwarf Laurel of Florida	
Andromeda Arborea	Sorrel Tree	b
——Plumata	Iron wood of Carolina	d
——Spicata	——	a
——Latifolia	Male Whortleberry	b
——Lanceolata	Virginia red Buds	a
——Nitida	Carolina red Bud	d
——Globulifera	Boggy Andromeda	
——Crassifolia	Evergreen Andromeda	a
Lonicera Canadensis	Canady Honey Suckle red and yellow	e
——Periclyminium	Virginia scarlet Honeysuckle	e
——Chamaecerassus	Dwarf Cherry	c
——Symphoricarpos	Indian Currants	b
Diarvilla Canadensis	Diarvilla	
Lythrum	Spiked Willow Herb	d
Xanthoxilum Virginianum	Tooth ach Tree	a
Glycine Frutescens	Kidney Bean Tree	
——Apios	Indian Patatoes	
Oxijacantha Canadensis	Canada Barberry	
Betula Papyrifera	White Birch of Canada	c
——Lenta	Red Birch	a
——Nigra	Sweet Birch	c
——Nana	D[illegible]arf Birch	a
——Populifolia	Aspen Birch	c
Alnus Rubra, Betula	Common Alder	d
——Maritima	Sea side Alder	a
——Glauca	Silver leaved Alder	
Acer Sacharissua	Sugar Maple	e
——Rubra	Scarlet Maple	a
——Glauca	Silver leav'd Maple	e
——Nigundo	Ash leav'd Maple	
——Arbustiva	Dwarf Mountain Maple	c
——Ornata	Striped Bark Maple	
Taxus Canadensis	Dwarf Yew	c
Mitchelia Repens	Creeping Syringa	Shade
Aristalochia Frutescens		
Tilia	Linden	e
Fraxinus Excelsior	Great White Ash Tree	a
——Nigra	Black Ash	c
Quercus Alba	. . .	o
——Nigra	. . .	b
——Rubra		
——Hispanica		
——Nana		
——Phyllos	Willow leaved Oak	a
——Deltoide	Water Oak	
——Folio Amplissima	Barren black Oak	f

Latin name	Common name
....	Shelbark Hycory
....	Great Shelbark H
.......Odorifera	Balsam Hicory
Corylus Nucleo Rotundiori and Duriori	Hazelnut
.....Cornata	Dwarf Filbert or c
Hamamelis	Witch Hazel
Ceanothus Foliis Trinerviis	Red Root
Celastrus Scandens	
Spirea Opulifolia	Nine Bark
.....Foliis Ternatis	Ipecacuanha
.....Aruncus	Spiraea (Indian
Rhus Vernix	Poison Ash
.....Cervina	Buck's Horn Shum
.....Lentisci Folia	Beech or Sea-side
.....Glabra	Scarlet Shumach
.....Canadensis	- -
.....Radicans	Poison Vine
.....Triphyllon	Poison Oak
Robinia Pseudacacia	Sweet flowering Lo
.....Villosa	Peach Blossom Aca
Gleditsia Triacanthus	Honey Locust
Bignonia Catalpa	Catalpa -
.....Crucigera	Cross Vine
.....Semper Virens	Yellow Jasmin
.....Radicans	Trumpet Flower
Nissa Tupilo	Water Tupilo
.....Ogeche	Ogeche Lime
......Sylvatica	Sower Gum
Aralia Spinoza	Prickly Ash
.....Nudicaulis	Spiknard False
.....Racemosa	Spiknard
Populus Delloidea	Cotton Tree
.....Foliis Cordatis	Black Poplar
.....Tremula	Aspen
Myrica Gale	Bog gale
.....Caerifera	Candleberry
.....Angustifolia	Dwarf Sweet Cand
Platanus Occidentalis	Button wood or wa
Grossularia Canadensis	Prickly Goosberry
Fagus Sylvatica	Beech Tree
.....Castanea	Chesnut
.....Castanea Pumila	Chinquapin
Calicanthus	Sweet Shrub of Ca
Ilex Aquifolium	Holley Tree
.....Yapon	Yapon or Cassena
.....Augustifolium	
.....Dahun	Dahun Holly
Gultheria	Jersey Tea
Epigea Procumbens	
Vitis Vinifera	Bunch Grape
....Labrusca	Small blue Grape
....Vulpina	Fox Grape red blac
Smilax Aspera	Bull Briar
....[illegible]nnua	Black Briony
	many more speci
Diosperos Guajacina	Persemmon
Æsculus Octandra	New River Horse C
....Caroliniana	Scarlet flowering ho
Dirca	Leather Bark
Clethra Alnifolia	Clethra

Cat